PRAISE FOR SEAN CARROLL

'Authoritative and beautifully written... Sean Carroll is a rare combination of excellent science writer and excellent research scientist. His writing exhibits a clarity of thought that is only available through a deep understanding of the subject.'
Brian Cox

'Weaving the threads of astronomy, physics, chemistry, biology and philosophy into a seamless narrative tapestry, Sean Carroll enthrals us with what we've figured out in the universe and humbles us with what we don't yet understand. Yet in the end, it's the meaning of it all that feeds your soul of curiosity.'
Neil deGrasse Tyson, host of Cosmos: A Spacetime Odyssey

'Like all great writers, Carroll has the remarkable ability of putting the reader utterly at ease with his lucid and addictive prose. He leads you so gently and comfortably into his quantum world that you quickly forget you are being given access to the most profound ideas about the nature of reality.'
Jim Al-Khalili, author of The World According to Physics

'Language, philosophy, quantum mechanics, general relativity – they're all in *The Big Picture*. Sean Carroll is a fantastically erudite and entertaining writer.'
Elizabeth Kolbert, author of The Sixth Extinction

'The science is authoritative, yet bold and lively. The narrative is richly documented, yet full of human drama. Carroll's saga pulls you aboard a modern voyage of discovery.'
Frank Wilczek,
Nobel Laureate and author of A Beautiful Question

'Sean Carroll's immensely enjoyable *Something Deeply Hidden* brings readers face to face with the fundamental quantum weirdness of the universe – or should I say universes? And by the end, you may catch yourself finding quantum weirdness not all that weird.'
Jordan Ellenberg, author of How Not To Be Wrong

'Carroll gives us a front-row seat to the development of a new vision of physics: one that connects our everyday experiences to a dizzying hall-of-mirrors universe in which our very sense of self is challenged... fascinating'
Katie Mack, author of The End of Everything

ALSO BY SEAN CARROLL

The Biggest Ideas in the Universe 1: Space, Time and Motion
Something Deeply Hidden
The Big Picture
The Particle at the End of the Universe
From Eternity to Here

THE
BIGGEST
IDEAS IN THE
UNIVERSE

QUANTA AND FIELDS

SEAN CARROLL

A Oneworld Book

First published in the United Kingdom, Republic of Ireland and Australia
by Oneworld Publications, 2024

Copyright © Sean Carroll, 2024

The moral right of Sean Carroll to be identified as the Author of this work has been asserted
by him in accordance with the Copyright, Designs, and Patents Act 1988

Illustrations by Sean Carroll unless otherwise noted.

All rights reserved
Copyright under Berne Convention
A CIP record for this title is available from the British Library

ISBN 978-0-86154-648-0
eISBN 978-0-86154-815-6

Book design by Tiffany Estreicher
Printed and bound in Great Britain by Clays Ltd, Elcograf S.p.A.

While the author has made every effort to provide accurate telephone numbers, internet
addresses, and other contact information at the time of publication, neither the publisher nor
the author assumes any responsibility for errors or for changes that occur after publication.
Further, the publisher does not have any control over and does not assume any responsibility
for author or third-party websites or their content.

Oneworld Publications
10 Bloomsbury Street
London WC1B 3SR
England

> Stay up to date with the latest books,
> special offers, and exclusive content from
> Oneworld with our newsletter
>
> Sign up on our website
> oneworld-publications.com

To Ariel

CONTENTS

	Introduction	1
ONE	Wave Functions	7
TWO	Measurement	33
THREE	Entanglement	57
FOUR	Fields	77
FIVE	Interactions	99
SIX	Effective Field Theory	123
SEVEN	Scale	149
EIGHT	Symmetry	167
NINE	Gauge Theory	193
TEN	Phases	213
ELEVEN	Matter	235
TWELVE	Atoms	253
	Appendix: Fourier Transforms	273
	Index	281

QUANTA AND FIELDS

INTRODUCTION

The history of physics has witnessed a number of brilliant, transformative ideas. But when it comes to truly revolutionary shifts—changes of paradigm that upend the way we think about the nature of reality—there have really only been two: classical mechanics in the late seventeenth century, and quantum mechanics in the early twentieth.

Classical mechanics was the theme of *The Biggest Ideas in the Universe: Space, Time, and Motion*, in which we emphasized the idea of the Newtonian/Laplacian paradigm, all the way up through spacetime and relativity. Now it's time for us to go quantum.

Quantum mechanics is, according to our best current understanding, the way the world works. The first hints of the need for a change came from work by Max Planck and Albert Einstein that indicated light was not merely a wave, as physicists had previously thought. In the right circumstances, light comes in particles we now call **photons**. These particles are an example of the **quanta** of the title—discrete bundles of energy emerging out of the rules of quantum mechanics. But it's subtler than that. Under different circumstances, things we think of as particles, like electrons and protons and neutrons, exhibit wave-like behavior. Quantum mechanics is going to continually frustrate our desire to put the behavior of physical systems into neat, commonsensical boxes.

Don't feel bad if the ideas of quantum mechanics seem alien at first. The truth is that physicists themselves don't agree on what, at rock bottom, quantum mechanics actually says. Physicists are extremely good at *using* quantum mechanics. We can predict the structure of atoms and molecules or calculate the scattering of particles off each other with exquisite precision. But it's a bit of a black box. The top quantum physicists in the world don't agree on what is going on to produce the results they predict and observe so successfully.

This lack of intellectual consensus can be traced to the fact that quantum mechanics seems to attribute special properties to the act of "measuring" or "observing" a physical system. In classical physics, objects have properties like positions and velocities, and you can directly measure them with as much accuracy as you like. Quantum objects seem profoundly different. Measuring the properties of a quantum system tends to dramatically change those properties. In some perfectly reasonable ways of thinking about the theory, a particle such as an electron doesn't even *have* properties like "position" or "momentum"—those are possible measurement outcomes, not intrinsic features of the quantum system itself.

For the most part, we're not going to worry about any of that.* Here at the Biggest Ideas in the Universe, our attitude is that of hard-nosed physicists, using well-established ideas to make testable predictions about the world. That will give us more than enough to chew on. The foundational issues are indisputably important; understanding

---

* If you're interested, I'm happy to recommend my book *Something Deeply Hidden* (Oneworld, 2019) for an introduction to the thorny problems at the heart of quantum mechanics. And for a fascinating history of the subject, check out Adam Becker's *What Is Real?* (John Murray, 2018).

INTRODUCTION

quantum mechanics at a deep level might very well turn out to be crucial to pushing beyond our current theories toward a much more comprehensive picture of reality. But the focus of this book will be on understanding the concepts underlying those current theories, and how they have given us an unprecedentedly accurate picture of the physical world.

The concepts in question include quantum mechanics itself; quantum field theory, which arises naturally when one combines quantum mechanics with the requirements of special relativity; and various deep ideas that have arisen within quantum field theory, including Feynman diagrams, renormalization, gauge theories, symmetry breaking, and the spin-statistics connection.

It's an enormous amount of material, which I've endeavored to boil down to its bare essence. The trick, as with the previous book, is that we are going to include enough mathematical specifics to understand the ideas for real, without reaching a level of detail required to solve problems in the manner of a graduate student studying for their doctorate. You will learn the same ideas that they will, but you won't have to pull all-nighters doing problem sets.

That requires a slightly different strategy than we employed in *Space, Time, and Motion*, although the basic aspiration is the same. In that book, I could literally show you all of the equations exactly as a professional would learn them. Here, there is just too much information to make that workable. Quantum field theory is loaded down with nitpicky details and layers of notation, which can get in the way of focusing on the central ideas. And it's the ideas that matter to us. So there will be times when we will ignore coupling constants, hide indices, treat matrices like numbers. I promise you that this is all in the service of helping you understand what's really going on, not in obscuring it.

Still, there is going to be math. In *Space, Time, and Motion* we introduced the basic ideas of calculus, including derivatives (rates of change) and integrals (accumulated amounts of change). In later chapters we dealt with tensors, and the use of Greek letters to denote spacetime indices. All of those are going to be here, in force. As well as the basic physics ideas of mass, energy, and relativity. If you are already familiar with those concepts, this book will be entirely self-contained. If not, *Space, Time, and Motion* should convey everything you need to know.

It will be a breathtaking ride. At the dawn of the twentieth century, classical mechanics was firmly entrenched. Twenty-five years later, we saw the first complete formulations of quantum mechanics. Twenty-five years after that, quantum electrodynamics was the first well-established quantum field theory. And twenty-five years after that, physicists had put together the Standard Model of particle physics, which remains triumphant to this day. That's the journey on which we are embarking, and it involves some of the most amazing ideas human beings have ever come across.

I have once again been extraordinarily fortunate to receive detailed feedback on drafts of this book. Enormous gratitude goes to Scott Aaronson, Justin Clarke-Doane, Ira Rothstein, and Matt Strassler, who kept my physics honest and my explanations not as convoluted as they would otherwise have been. My agent, Katinka Matson, has provided sage advice along the way. And huge thanks to my editor, Stephen Morrow, who has been patient and understanding and singularly helpful in shaping this book series into something I hope people will learn from and enjoy.

Patience was especially called for this time around, as the writing process was interrupted by a cross-country move and beginning a position at Johns Hopkins. Thanks to all my new colleagues and students for making everything as smooth as possible and being understanding when I wasn't always as available as I might have liked.

Most of all, thanks to my wife, Jennifer, who picked up and moved with me, shouldered most of the burden of shaping our new home, and always keeps my writing honest. Looking forward to this new and exciting chapter.

The plot of the Mexican-hat potential in Chapter 10 is adapted from a Mathematica code by Vitaliy Kaurov (https://mathematica.stackexchange.com/questions/19578/how-can-i-make-a-plot-of-the-higgs-potential). The image of the LIGO observatory in Hanford in Chapter 11 is from the LIGO collaboration (https://www.ligo.org/multimedia/gallery/ lho-images/Aerial5.jpg). The plot of nuclides in Chapter 12 is based on an example from *Learning Scientific Programming with Python* by Christian Hill (https://scipython.com/).

ONE

WAVE FUNCTIONS

As the nineteenth century drew to a close, you would have forgiven physicists for hoping that they were on track to understand everything. The universe, according to this tentative picture, was made of particles that were pushed around by fields.

The idea of **fields** filling space had really taken off over the course of the 1800s. Earlier, Isaac Newton had presented a beautiful and compelling theory of motion and gravity, and Pierre-Simon Laplace had shown how we could reformulate that theory in terms of a gravitational field stretching between every object in the universe. A field is just something that has a value at each point in space. The value could be a simple number, or it could be a vector or something more complicated, but any field exists everywhere through space.

But if all you cared about was gravity, the field seemed optional—a point of view you could choose to take or not, depending on your preferences. It was equally okay to think as Newton did, directly in terms of the force created on one object by the gravitational pull of others without anything stretching between them.

That changed in the nineteenth century, as physicists came to grips with electricity and magnetism. Electrically charged objects exert forces on each other, which is natural to attribute to the existence of an electric field stretching between them. Experiments by Michael Faraday showed that a moving magnet could induce electrical current in a wire without actually touching it, pointing to the existence of a separate magnetic field, and James Clerk Maxwell managed to combine these two kinds of fields into single a theory of **electromagnetism**, published in 1873. This was an enormous triumph of unification, explaining a diverse set of electrical and magnetic phenomena in a compact theory. "Maxwell's equations" bedevil undergraduate physics students to this very day.

One of the triumphant implications of Maxwell's theory was an understanding of the nature of **light**. Rather than a distinct kind of substance, light is a propagating **wave** in the electric and magnetic fields, also known as **electromagnetic radiation**. We think of electromagnetism as a "force," and it is, but Maxwell taught us that fields carrying forces can vibrate, and in the case of electric and magnetic fields those vibrations are what we perceive as light. The quanta of light are particles called photons, so we will sometimes say "photons carry the electromagnetic force." But at the moment we're still thinking classically.

Take a single charged particle, like an electron. Left sitting by itself, it will have an electric field surrounding it, with lines of force pointing toward the electron. The force will fall off as an inverse-square law, just as in Newtonian gravity.* If we move the electron, two things happen: First, a charge in motion creates a magnetic field as well as an electric one. Second, the existing electric field will adjust how it is oriented in space, so that it remains pointing toward the

---

* Later in this book we will actually see why both gravity and electromagnetism have inverse-square laws. It is because the underlying fields are massless, and that in turn is because of something called gauge invariance.

particle. And together, these two effects (small magnetic field, small deviation in the existing electric field) ripple outward, like waves from a pebble thrown into a pond. Maxwell found that the speed of these ripples is precisely the speed of light—because it *is* light. Light, of any wavelength from radio to x-rays and gamma rays, is a propagating vibration in the electric and magnetic fields. Almost all the light you see around you right now has its origin in a charged particle being jiggled somewhere, whether it's in the filament of a lightbulb or the surface of the sun.

Simultaneously in the nineteenth century, the role of **particles** was also becoming clear. Chemists, led by John Dalton, championed the idea that matter was made of individual **atoms**, with one specific kind of atom associated with each chemical element. Physicists belatedly caught on, once they realized that thinking of gases as collections of bouncing atoms could explain things like temperature, pressure, and entropy.

But the term "atom," borrowed from the ancient Greek idea of an indivisible elementary unit of matter, turned out to be a bit premature. Though they are the building blocks of chemical elements, modern-day atoms are not indivisible. A quick-and-dirty overview, with details to be filled in later: atoms consist of a **nucleus** made of **protons** and **neutrons**, surrounded by orbiting **electrons**. Protons have positive electrical charge, neutrons have zero charge, and electrons have negative charge. We can make a neutral atom if we have equal numbers of protons and electrons, since their electrical charges will cancel each other out. Protons and neutrons have roughly the same mass, with neutrons being just a bit heavier, but electrons are much lighter, about $1/1,800$th the mass of a proton. So most of the mass in a person or another macroscopic object comes from the protons and neutrons. The lightweight electrons are more able to move around and are therefore responsible for chemical reactions as well as the flow of electricity. These days we know that protons and neutrons are themselves made of smaller particles called **quarks**, which are held together by **gluons**, but there was no hint of that in the early 1900s.

This picture of atoms was put together gradually. Electrons were discovered in 1897 by British physicist J. J. Thompson, who measured their charge and established that they were much lighter than atoms. So somehow there must be two components in an atom: the lightweight, negatively charged electrons, and a heavier, positively charged piece. A few years later Thompson suggested a picture in which tiny electrons floated within a larger, positively charged volume. This came to be called the plum pudding model, with electrons playing the role of the plums.

The plum pudding model didn't flourish for long. A famous experiment by Ernest Rutherford, Hans Geiger, and Ernest Marsden shot alpha particles (now known to be nuclei of helium atoms) at a thin sheet of gold foil. The expectation was that they would mostly pass right through, with their trajectories slightly deflected if they happened to pass through an atom and interact with the electrons (the plums) or the diffuse positively charged blob (the pudding). Electrons are too light to disturb the alpha particles' trajectories, and a spread-out positive charge would be too diffuse to have much effect. But what happened was, while most of the particles did indeed zip through unaffected, some bounced off at wild angles, even straight back. That could only happen if there was something heavy and substantial for the particles to carom off of. In 1911 Rutherford correctly explained this result by positing that the positive charge was concentrated in a massive central nucleus. When an incoming alpha particle was lucky enough to score a direct hit on the small but heavy nucleus, it would be deflected at a sharp angle, which is what was observed. In 1920 Rutherford proposed the existence of protons (which were just hydrogen nuclei, so had already been discovered), and in 1921 he theorized the existence of neutrons (which were eventually discovered in 1932).

So far, so good, thinks our imagined fin de siècle physicist. Matter is made of particles, the particles interact via forces, and those forces are carried by fields. The entire mechanism would run according to rules established by the framework of classical physics. For particles this is pretty familiar: we specify the positions and the momenta of all the particles, then use one of our classical techniques (Newton's laws or their equivalent) to describe their dynamics. Fields work in essentially the same way, except that the "position" of a field is its value at every point in space, and its "momentum" is how fast it's changing at every point. The overall classical picture applies in either case.

The notion that physics was close to being all figured out was tempting. Albert Michelson, at the dedication of a new physics laboratory at the University of Chicago in 1894, proclaimed, "It seems probable that most of the grand underlying principles [of physics] have been firmly established."

He was quite wrong.

But he was also in the minority. Other physicists, starting with Maxwell himself, recognized that the known behavior of collections of particles and waves didn't always accord with our classical expectations. William Thomson, Lord Kelvin, is often the victim of a misattributed quote: "There is nothing new to be discovered in physics now. All that remains is more and more precise measurement." His real view was the opposite. In a lecture in 1900, Thomson highlighted the presence of two "clouds" looming over physics, one of which was eventually to be dispersed by the formulation of the theory of relativity, the other by the theory of quantum mechanics.

BLACKBODY RADIATION

The history of science is subtle and complicated, and progress rarely takes the straight path we remember in retrospect. Quantum mechanics in particular had a painful and messy development. We're going to skip over many of the historical twists and turns to focus on two puzzling phenomena that kicked off the quantum revolution: waves exhibiting particle-like properties, and particles exhibiting wave-like properties.

The particle-like properties of light came first. The idea arose from studying **blackbody** (or "thermal") **radiation**, which is the radiation emitted by an object that absorbs any incident light but nevertheless radiates just because it has a nonzero temperature. To physicists, the temperature of an object characterizes the random jiggling of its constituent particles, and randomly jiggling particles are going to emit radiation depending on how fast they are moving. When you look at a painting, you see an intricate configuration of shape and color that reflects the light that is shining on it. A blackbody, by contrast, is what you get when you turn off all the ambient light and just let objects glow because of their temperature; the glow from a heating element on an electric stove is a good example. Everything with a nonzero temperature gives off some thermal radiation, but pure blackbody radiation depends only on the temperature, unspoiled by color or reflectivity or other properties of the object. A low-temperature blackbody will primarily radiate at infrared or even radio wavelengths, and as we increase the temperature we see more visible light, ultraviolet, and ultimately x-rays.

So blackbody radiation represents a seemingly simple physics problem (a spherical cow, one might say). It has a temperature, and none of its other properties matter. Temperature measures the kinetic energy of atoms in the body jiggling back and forth, and those atoms contain charged particles, so this jiggling leads to the emission of electromagnetic radiation. Our physics problem is, how much radiation is given off at each wavelength?

Physicists in the nineteenth century set about both measuring the radiation as a function of wavelength—the **spectrum** of the blackbody—and calculating it theoretically. The measured curve is a thing of beauty, climbing up from zero at short wavelengths to a peak that depends on the temperature, then decaying back down to zero at long wavelengths.

The theoretical situation, however, was a mess. One proposed theory, by Wilhelm Wien in 1896, seemed to fit well at short wavelengths but diverged from experimental data at longer wavelengths. Another, by John Strutt (Lord Rayleigh) in 1900, worked the other way around: his fit well at long wavelengths but not at short ones. Indeed, it predicted an infinite amount of radiation at short wavelengths.

Rayleigh's calculation, later improved upon by James Jeans, is generally considered to more accurately reflect what we would expect to observe if the world had actually been classical. Its failure at short wavelengths has been dubbed an **ultraviolet catastrophe**, as physicists puckishly refer to anything that happens at short distances as ultraviolet or just "UV," and anything that happens at long distances as infrared or "IR." (And any mismatch between theory and experiment as a catastrophe.) The relative ease of understanding IR phenomena, and the relative difficulty of getting the UV right, will come back with a vengeance when we reach quantum field theory.

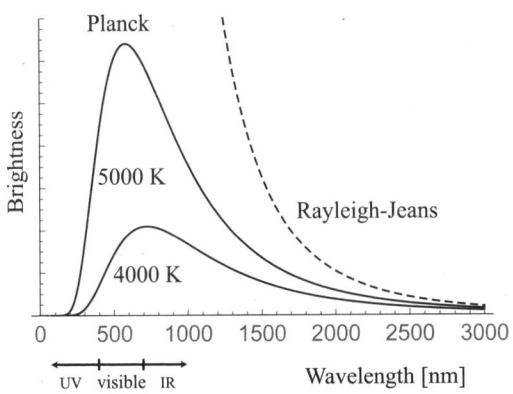

German physicist Max Planck—who, as the story goes, was told by one of his professors not to go into physics, because "almost everything is already discovered"—decided to tackle the problem. In 1900 he was able to write down a formula that was a compromise between Wien and Rayleigh-Jeans, fitting the observations at both long wavelengths and short ones (as well as in between). His result, the famous **Planck blackbody radiation law**, gives us the brightness B of an object at temperature T at each wavelength λ:

$$B(\lambda) = \frac{2hc^2}{\lambda^5} \frac{1}{\exp\left(\frac{hc}{\lambda k_B T}\right) - 1}. \tag{1.1}$$

Here "exp" stands for the exponential function, $\exp(x) = e^x$. In addition to the temperature T and wavelength λ, this expression depends on three fixed parameters: the speed of light c, Boltzmann's constant k_B from thermodynamics, and a new constant h that Planck had to invent to complete the formula. Now known as **Planck's constant**, this number shows up everywhere quantum mechanics is relevant:

$$h = 6.626 \times 10^{-34} \text{ Joules} \cdot \text{second}. \tag{1.2}$$

A joule is a unit of energy, equivalent to the amount a one-watt lightbulb uses in one second. For various reasons it turns out that h appears frequently divided by 2π, so that we define the **reduced Planck constant** as

$$\hbar = \frac{h}{2\pi}, \tag{1.3}$$

and pronounce it as "h-bar." Soon enough we'll realize that this constant is so ubiquitous that we tend to choose units where $\hbar = 1$, just as it is convenient to set the speed of light to $c = 1$ when we work with relativity. But for now let's keep it around.

At first, Planck didn't so much derive his formula as guess it. He worked out the right mathematical manipulations that would combine the Wien and Rayleigh-Jeans results into a single compact expression. But he worked hard to come up with a reason why such a formula should work so well. Part of why the task was challenging was that Planck was, at heart, a conservative physicist. He wasn't fond of statistical mechanics à la Maxwell and Ludwig Boltzmann, which purported to explain the laws of thermodynamics in terms of the collective behavior of large numbers of atoms, and was even skeptical about the very existence of atoms themselves. Later in life he remained dubious of the main ideas of quantum mechanics. But that didn't stop him from hitting on an ingenious assumption that solved the puzzle of blackbody radiation.

The crucial conjecture was this: when a charged particle in a body jiggles back and forth and emits electromagnetic radiation, the amount of energy contained in that radiation cannot be any old number. Instead, it is emitted in discrete amounts depending on the frequency of the wave, which is related to its wavelength and the speed of light by $f = c/\lambda$. Higher-frequency light waves correspond to higher-energy bundles, according to a now-famous formula,

$$E = hf, \qquad (1.4)$$

where we once again see the appearance of Planck's constant. As it is often more convenient to use the angular frequency $\omega = 2\pi f$ rather than the frequency itself, this equation is often written as $E = \hbar\omega$.

WAVES ACTING LIKE PARTICLES

Why would the energy of emitted radiation come in discrete chunks, rather than being allowed to have any value at all? Perhaps because the particles in the blackbody jiggle only in discrete amounts. But an alternative explanation springs to mind: not that the jiggling is discrete but that the emitted light is. In other words, that light actually is a stream of discrete entities—particles.

But Planck didn't go so far as to say that; his statement was only about the amount of energy emitted, not about what form it took. Whether light is a particle or a wave is an old question, going back to at least Isaac Newton (who advocated for particles) versus Christiaan Huygens (who defended waves). Once Maxwell's equations came on the scene, physicists were pretty convinced that light was a wave. In particular, experiments had demonstrated that light interferes with itself, as we expect when the positive part of a wave cancels against the negative part. Not to mention that Maxwell had explained the nature of light in terms of his theory of electromagnetism, which was successful for numerous other reasons. So the idea that there was something particle-like about light seemed to be an unlikely suggestion; physicists assumed they had moved on from that concept.

The person with the gumption to make the leap was a twenty-six-year-old Albert Einstein; 1905 has come to be known as Einstein's annus mirabilis, or "miraculous year." In a series of papers, he formulated special relativity, articulated the relationship between mass and energy, and explained Brownian motion (the random motion of microscopic particles in liquid) in terms of atomic collisions, which helped convince scientists of the existence of atoms once and for all. Any one of these achievements would have made the career of an ordinary scientist, but Einstein didn't even win the Nobel Prize for any of them.

The one for which he did win the Nobel proposed the idea of **energy quanta** of light. "Quanta" is the plural of "quantum," meaning "a smallest portion into which something can be subdivided." The word **photon** wasn't coined until later, but that's precisely what Einstein suggested: that light consists of particles, now known as photons.

He did this in order to explain a somewhat-obscure phenomenon known as the **photoelectric effect**. When you shine light at a metal, it will sometimes kick out an energetic electron. The effect doesn't depend on the brightness or intensity of the light, as we might expect if light is a continuous wave of energy, but only on its frequency. That makes sense if instead light comes in quantum packets (photons) with individual energies given by equation (1.4). When the energy of a photon is enough to knock an electron loose, it does; when it isn't, it doesn't matter how many photons you send, the electrons will remain safely in place. This picture also suffices, as Einstein noted, to explain Planck's formula for blackbody radiation.

However, it flew in the face of everything physicists thought they had learned about light. The picture of light as a wave hadn't been settled on casually; it was backed by strong experimental evidence and theoretical reasoning. You couldn't just say "light is particles, after all" and be done with it. It seemed, rather, that light behaved in wave-like ways much of the time but had particle-like properties in certain special situations. If that seems messy and vague, that's because it is. It would take another two decades of flailing around before something like a coherent explanation would begin to emerge. (And even now, a century after that, there is no consensus on what is truly going on.)

PARTICLES ACTING LIKE WAVES

Meanwhile, Ernest Rutherford and his associates, most prominently Danish physicist Niels Bohr, were working to understand the atomic structure of matter. It was Rutherford's experiments in 1911 that had established that most of the mass in an atom was concentrated in a dense, positively charged nucleus at the center. The question was what happened with the much lighter electrons.

An obvious idea would be that the electrons orbited the nucleus, much like planets orbit the sun. Something along these lines had previously been suggested by Irish physicist Joseph Larmor, and an alternative framework in which electrons moved in rings (like those of the planet Saturn) was put forward by Japanese physicist Hantaro Nagaoka. Rutherford himself was vague about what the electrons were doing, but he knew they had to be moving around the nucleus somehow. And he knew roughly the size of the atoms as a whole, and therefore the orbits of the electrons.

And this, people soon recognized, created an enormous problem. As we just discussed, charged particles in motion give off electromagnetic radiation. That includes the motion of electrons orbiting an atomic nucleus, at least according to the rules of classical mechanics. So our electrons should give off light and in doing so lose energy. As a result, they shouldn't peacefully stay in orbit around the nucleus. An orbiting electron would quickly plunge right into the nucleus itself, shedding the energy it loses in a stream of electromagnetic waves. You can even calculate how long this should take, with an answer of about 10^{-11} seconds. But if all the atoms in the universe lasted only that long, someone would have noticed by now.

According to the rules of classical mechanics, in other words, all matter made from atoms should be dramatically unstable. Tables and chairs and planets and people should collapse into specks within a tiny fraction of a second. But that doesn't seem to happen. Why?

An initial answer was proposed by Bohr in 1913. People knew about the ideas of Planck and Einstein concerning radiation, so the notion of "quantum" was already in the air. Bohr suggested that there was something vaguely quantum about electrons and their motion, just as there was for photons. Of course, electrons were already thought of as particles, but Bohr's proposal was that they could be attached to atoms in only certain discrete orbits, rather than at any old distance. Electrons are already "quanta," just because they are particles, but Bohr was saying that their allowed orbits are quantized as well.

The idea worked pretty well, at least for hydrogen, the simplest atom. Best of all, in order to fit the data, the quantization condition that Bohr had in mind amounted to an insistence that the angular momentum L of an orbiting electron would have to have the value

$$L = n\hbar, \qquad (1.5)$$

where n is any integer greater than 0. This was a seemingly miraculous appearance of Planck's constant, initially proposed to understand blackbody radiation, in a formula for the orbits of electrons around atomic nuclei. And the insistence on a fixed set of orbits solved the stability problem: electrons could at best decay to the lowest-energy orbit, not all the way to the nucleus. There is no allowed lower-energy state for them to go to, which suggested that atoms should be stable once all of their electrons are in their lowest-energy orbits.

The Bohr model was an important advance, but it lacked any explanation for *why* the electrons should be so picky about their orbits. An explanation was finally put forward in 1924 by French physicist Louis de Broglie as part of his doctoral thesis. In retrospect it's a simple idea: if light has particle-like properties, shouldn't we imagine that particles have wave-like properties as well?

The question remained what it might mean for something so manifestly particle-like as an electron to "have wave-like properties." It seems de Broglie had the sense that electrons consisted of both particles and waves, with the particles following the waves around. But the important idea was to imagine that the waves have a corresponding wavelength, which could be related to the momentum of the particle, $p = mv$. Thus the **de Broglie wavelength** is

$$\lambda = h/p, \qquad (1.6)$$

where h is once again Planck's constant. This wavelength could be used, in de Broglie's picture, to anticipate wave-like phenomena, such as constructive or destructive interference when two "matter waves" came together.

Most important, de Broglie's model suggested a natural explanation for Bohr's quantized electron orbits. To wit: the orbits have to be exactly the right size to fit an integer number of wavelengths on them. In other words, the wave has to come back to the same value it started when it goes once around the orbit. That condition turns out to be just right to reproduce the orbits described by the Bohr model, strongly suggesting that this idea was on the right track.

QUANTUM MECHANICS

But Einstein's light quanta, Bohr's discrete orbits, and de Broglie's matter waves still felt more like a grab bag of ideas than a fully fleshed-out theory. The first complete and rigorous formulation of what we now call **quantum mechanics** was put forward in 1925 by a trio of German physicists: Werner Heisenberg, Max Born, and Pascual Jordan. The basic idea, now known as **matrix mechanics**, came first to Heisenberg while he was trying to recover from hay fever by convalescing on the island of Helgoland. The idea was both simple and revolutionary: if electron orbits are so troublesome, let's deny that there are any such things as "electron orbits." Forget about what's "really happening" and focus only on what we can *observe*.

Heisenberg proposed that momentum and position should be thought of as **observables**—not quantities with definite values but questions we can ask by doing measurements. The question "What is the position of the electron?" doesn't have an answer before you ask it. You can "measure the position or momentum," but by doing so you are bringing the measurement's outcome into existence, not revealing a preexisting truth. With this insight Heisenberg was able to correctly derive the way that light was emitted from atoms.

Heisenberg was still young at the time (twenty-three years old), and he fretted that his model was a little too audacious. He wrote up

WAVE FUNCTIONS

an article articulating his idea, but before submitting it he sent it to his senior colleague Max Born, warning that he "had written a crazy paper." Born had the right mathematical training to recognize that Heisenberg's model was best expressed in the language of **matrices**—replacing single numerical quantities with square arrays of numbers. Born and his former student Pascual Jordan wrote a follow-up paper of their own. Then all three of them collaborated on yet another paper, fleshing out the details.

The problem was that, just as Heisenberg wasn't familiar with the idea of matrices before Born connected the dots, most other physicists weren't either. The mathematical formalism seemed solid enough, but the underlying physical meaning was obscure, and there was some reluctance in the physics community to declare victory just yet.

Soon thereafter, however, Austrian physicist Erwin Schrödinger came up with a seemingly different approach. Schrödinger, following de Broglie, put waves once again at the center of the story, and his theory was dubbed **wave mechanics**. Eventually it was shown that matrix mechanics and wave mechanics are two equivalent ways of representing the same physical theory, so nowadays we simply say **quantum mechanics**.

To extend the idea of de Broglie's matter waves, Schrödinger proposed what we now call the **wave function**, often written $\Psi(x)$, where Ψ is the capital Greek letter psi—such an unromantic name for something that would come to hold central importance to the fundamental nature of reality. Just as giving the position and momentum of every constituent specifies the classical state of a system, giving the wave function specifies the **quantum state** of a system. If we're considering just a single particle, its wave function assigns a number to every spatial location, just as for any other kind of wave. When we have more than one particle, things aren't so simple, because the wave function is *not* just any old wave. This is because of **entanglement**, which we'll discuss to death in later chapters.

What does the wave function physically represent? Well, that's a good question. Schrödinger originally thought of it as something fairly tangible, like the density of matter. But it was ultimately reinterpreted as a way of calculating probabilities of measurement outcomes, as we'll see. Let's put this crucial question aside for the moment.

A big difference between de Broglie's matter wave and Schrödinger's wave function is that the wave function at any one point is a **complex number**, formed from combining a real number and an imaginary number:

$$\Psi(x) = \Psi_R(x) + i\,\Psi_I(x). \tag{1.7}$$

Here, $i = \sqrt{-1}$ is the "imaginary unit." The functions $\Psi_R(x)$ and $\Psi_I(x)$ are the real and imaginary parts, respectively, of $\Psi(x)$. Note that both $\Psi_R(x)$ and $\Psi_I(x)$ are themselves real numbers; you multiply a real number by i to get an imaginary number, which is what's happening in the second term on the right-hand side of (1.7). You can think of the real part and the imaginary part as two axes in a **complex plane**, illustrated in the figure below.

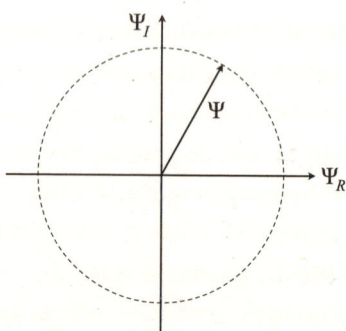

Why is the wave function complex instead of real? Ultimately it's because that's how nature works, rather than some other way. But the complex nature of $\Psi(x)$ does allow for a nice feature: the overall *shape*

of the wave function (of, for example, an electron in an atom) can remain fixed, while the function nevertheless evolves, just by trading off its real part for its imaginary part and vice versa. In the figure (where we're just illustrating the value of Ψ at some particular x, rather than the whole function), this corresponds to rotating Ψ around in a circle while keeping its length fixed.

It does, however, make the wave function seem a bit abstract. Something like an electric field seems relatively concrete. The electric field has a value, which is a little vector, at every point in space. We can even measure that value by placing a charged particle there and watching the electric field push it around. Schrödinger's wave function doesn't seem quite as *real*, literally as well as mathematically. How did complex numbers sneak into our best description of reality? You may wonder, what kind of thing can you measure to get an imaginary-number result? But physicists are more than willing to brush aside such nagging issues if it means they can have a simple theory that makes testable predictions, which quantum mechanics surely does.

THE SCHRÖDINGER EQUATION

What really made the wave function an instant hit among physicists, however, was the presence of a dynamical equation governing how it changes over time. Now known as the **Schrödinger equation**, it takes different specific forms depending on what system is being described. (Just like we can plug different forces into Newton's second law, $\vec{F} = m\vec{a}$.) In its most general, abstract form, the Schrödinger equation looks like this:

$$\widehat{H}\Psi = i\hbar \frac{\partial \Psi}{\partial t}. \qquad (1.8)$$

This might look a bit scary, but that's only because the symbols are unfamiliar. The equation is actually extremely simple. We'll walk through some of the details, but at the end of the day all you have to remember is that the Schrödinger equation tells us how the wave function evolves over time.

We see once again the appearance of the imaginary unit i and the reduced Planck constant $\hbar$. The right-hand side is just the partial derivative of the wave function Ψ with respect to time t. Remember that a partial derivative is a way of saying "keep everything else fixed, and calculate the rate of change with respect to t." This is what makes Schrödinger's equation, unlike de Broglie's relation (1.6), a *dynamical* relation: you specify the wave function at one moment in time, and the equation determines what it will be at the next moment, as well as all subsequent moments. The Schrödinger equation fits in perfectly with the Laplacian paradigm of classical mechanics, in which the information that specifies the state is conserved as the system evolves. (As we'll see, the wave function seems to evolve in an entirely different way when the system is measured rather than being left alone—that's the source of all the mystery of quantum mechanics.)

The tricky part is the left-hand side, $\widehat{H}\Psi$. The letter H stands for the **Hamiltonian**, familiar from our investigation of classical mechanics in *Space, Time, and Motion*. There, we had position x and momentum p as coordinates on phase space, and the Hamiltonian $H(x, p)$ was simply the energy of the system written as a function of those coordinates.

The quantum situation is trickier, but in a fun way. The Hamiltonian is no longer a function of phase space but rather an **operator**. We put a hat on it and write it as $\widehat{H}$ to remind ourselves of that. By an "operator" we mean a mathematical procedure that takes in a function and spits out another function. To be pedantic about it, the Hamiltonian operator is a map from the original function Ψ to a new function, denoted $\widehat{H}\Psi$:

$$\widehat{H} : \Psi(x) \to \widehat{H}\Psi(x). \tag{1.9}$$

WAVE FUNCTIONS

In classical mechanics, the Hamiltonian is just the energy; in quantum mechanics, it's an operator that breaks the wave function into pieces, asks "how much energy is there in this piece?," then adds the results together to get a new function. It's good that the Hamiltonian acting on the wave function gives us another function rather than simply a number, since we want to set it proportional to $\partial \Psi / \partial t$, which is certainly a function of x itself. (The wave function, and the function we get by acting the Hamiltonian on it, also depend on time, but to keep things simple we need not write that out explicitly right now.)

So this is how we are to think about quantum states and their evolution, in Schrödinger's picture. We have a system, like a single particle moving in a one-dimensional potential, and it has some coordinates, like the single number x. We start at some initial time with a complex-valued wave function that depends on the coordinates, $\Psi(x)$. We then invoke a Hamiltonian operator, $\widehat{H}$, which acts on the wave function to give a new function. What Hamiltonian we actually use just depends on what kind of system we're considering, and in particular what kind of energy it has. And that new function tells us $i\hbar$ times the derivative of Ψ with respect to time. The rate at which the wave function evolves depends on the energy of the quantum state; energetic states evolve more rapidly, and lower-energy states more slowly.

ONE PARTICLE

Enough with the abstract nonsense. Let's bring it down to Earth by thinking about what the Hamiltonian operator actually does.

That turns out to be a heavy lift. Every system is described by some Hamiltonian, whether it's a single particle, the Standard Model of particle physics, or the universe as a whole. But the Hamiltonian will be different in each case. Much of the work of a theoretical physicist is deciding what the right Hamiltonian should be for a system, since that choice governs the dynamics of the system. From the Schrödinger perspective, "choosing a Hamiltonian" is equivalent to "choosing the laws of physics." It's analogous to "choosing what forces act on a system" in classical mechanics.

Schrödinger himself wasn't thinking so abstractly—at least, not at first. He was, quite reasonably, starting with a simple system: a single particle of mass m moving along one dimension in a potential $V(x)$. Literally a ball rolling on a hilly landscape, although conceived of quantum-mechanically. But let's stick to non-relativistic particles (moving slowly compared to the speed of light), since the combination of relativity with quantum mechanics turns out to require quantum field theory to do it right.

The nice thing is that we can generally—most of the time anyway—work out what the quantum Hamiltonian operator for a system should be if we start with the good old classical Hamiltonian. For a non-relativistic particle, we know what the classical Hamiltonian would be. It's the kinetic energy (written in terms of the momentum) plus the potential energy, so

$$H(x,p) = \frac{p^2}{2m} + V(x). \tag{1.10}$$

To convert this into a quantum Hamiltonian operator, we can go piece by piece, treating both x and p as operators themselves, which we now denote $\hat{x}$ and $\hat{p}$. The operation of $\hat{x}$ is just "multiply by x," which isn't so hard. The tricky part is the momentum $\hat{p}$. In classical mechanics, that was an independent variable that helped to define the state. For any x, we were welcome to consider states with any p.

This is no longer true in quantum mechanics. The wave function $\Psi(x)$ defines the entire quantum state; there is no extra dependence on p. Instead, momentum is now an operator, which is proportional to the partial derivative of Ψ with respect to x:

$$\hat{p} = -i\hbar \frac{\partial}{\partial x}. \tag{1.11}$$

You don't have to fret too much about where this comes from or why it's true. For the moment just appreciate that momentum is related to the spatial derivative of the wave function, which is just the slope of the curve. A gently meandering Ψ is characterized by low momentum, while a sharply oscillating Ψ will have high momentum.

Now all we have to do is plug (1.11) into the classical expression for the Hamiltonian (1.10) to get a formula for the quantum Hamiltonian. And that gives us the Schrödinger equation in the form in which Schrödinger originally wrote it down:

$$\left(-\frac{\hbar^2}{2m}\frac{\partial^2}{\partial x^2}+V(x)\right)\Psi(x,t)=i\hbar\frac{\partial}{\partial t}\Psi(x,t). \quad (1.12)$$

The abstract form of Schrödinger's equation, (1.8), has the kind of austere beauty we would want from a fundamental law of physics. The explicit form (1.12) is a bit clunkier, but we have to be detailed if we want to make experimental predictions. Each year, in universities around the world, countless young physics students stay up all night solving this equation for various physical situations.

THE SIMPLE HARMONIC OSCILLATOR

Happily we won't have to spend time explicitly solving anything. We're less interested in specific solutions to the Schrödinger equation and more interested in the general principle: there is a wave function, and it obeys a definite dynamical equation, which sets its rate of change proportional to the energy of (really, the Hamiltonian operator acting on) the wave function.

Still, a look at a specific kind of solution might be good for the soul. And what better example to consider than our old friend the simple harmonic oscillator, which we encountered in *Space, Time, and Motion*? This example isn't chosen solely for the sake of simplicity; it will turn out to be crucially important once we get to quantum field theory. We will start with smooth functions and see the emergence of "quanta," which will relate directly to why quantized fields look like particles.

Remember that the simple harmonic oscillator is defined by a potential energy that is quadratic in the coordinate. We can write it as

$$V(x) = \frac{1}{2} m \omega^2 x^2, \qquad (1.13)$$

where m is still the mass of the particle, ω is the angular frequency of the oscillator, and x is the coordinate. We are tempted to say "x is the position," but that would be your classical intuition talking. Position is an observable, and we might get any particular answer x were we to measure it, but before we make the measurement there's no such thing as "the position of the particle."

Of course, there are an infinite number of solutions to Schrödinger's equation in the harmonic-oscillator potential, since we can just start with whatever $\Psi(x)$ we fancy and then use the equation to determine how it evolves over time. (The Laplacian paradigm, familiar from classical physics.) But there are certain solutions that are especially interesting: ones that keep their shape fixed as time passes. These are also solutions for which the energy has a definite value rather than something uncertain. They are therefore called **energy eigenstates**. There is a lowest-energy state, called the **ground state**, and then **excited states** with the next-highest energy, and the next-highest after that, and so on. If we label them by $n = \{0, 1, 2, \ldots\}$, the energies of these states take the form

$$E_n = \hbar \omega \left(n + \frac{1}{2} \right), \qquad (1.14)$$

where ω is the angular frequency of the oscillator.

The precise mathematical form of energy eigenstates as functions of x turns out to be complicated and not especially illuminating.

WAVE FUNCTIONS

Fortunately, we can wave our hands and draw some pictures to get an intuitive idea of what is going on.

The harmonic-oscillator Hamiltonian depends on both the potential energy, which is lowest near $x = 0$, and the kinetic energy, which is lowest when $\Psi(x)$ is a gentle curve rather than wildly oscillating. And crucially, the potential energy becomes arbitrarily large at large values of x, since it grows as x^2. So any finite-energy state is going to have to approach zero at $x = -\infty$ and $x = +\infty$. If the wave function approaches any fixed nonzero value at $x = -\infty$, that would cost infinite energy, since it's multiplying a growing potential in the Schrödinger equation (1.12).

So, traveling from $x = -\infty$ to $x = +\infty$, we can more or less guess what the lowest-energy wave function ($n = 0$) is going to look like. It will start at zero when $x = -\infty$, but it can't just stay zero, otherwise there's no particle anywhere at all. So as we approach $x = 0$, the wave function is going to gently rise, then start falling back to zero again. It will never cross $\Psi = 0$, since we're thinking of the lowest-energy state, and that is going to vary as gently as possible, no extraneous wiggles.

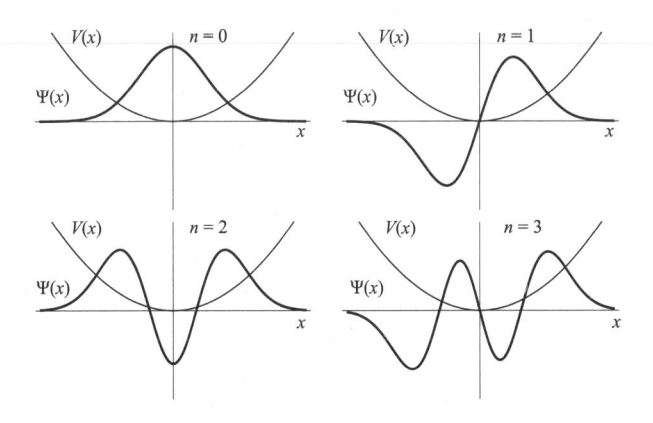

The next-highest energy ($n = 1$) will be spread out a little more and have a little more wiggle to it. So it will cross $\Psi = 0$ exactly once:

starting at zero, going down rather than up, coming back to $\Psi = 0$ at $x = 0$, then rising and falling again back to zero. The next highest after that, $(n = 2)$, will cross twice: starting at zero, going up, then back down across the axis, then back up again, and falling back to zero. The pattern repeats: each energy eigenstate labeled n will vary up and down, crossing the $\Psi = 0$ axis exactly n times.

Amid these mathematical details, it's important not to miss that a bit of a miracle has occurred here. We started our journey with the observation by Planck and Einstein that there was something discrete, or "quantum," in the behavior of photons, followed by Bohr's application of an analogous idea to electron orbits. But there's nothing discrete or quantum about wave functions or the Schrödinger equation. The wave function itself is perfectly smooth, as is its evolution over time.

The simple harmonic oscillator reconciles this apparent tension: it's not the wave function or the equation that it obeys that is discrete, it's some particular set of *solutions* to that equation that has a discrete character. That's where quanta come from.

Whenever wave functions are pinned down at infinity, they will oscillate around zero some discrete number of times. It's precisely analogous to the strings on a violin or a guitar, which are literally tied down at each end. If we pluck such a string, it can vibrate at the longest possible wavelength, corresponding to the fundamental frequency, or at discrete higher frequencies, corresponding to overtones. The mathematical formula will be different for wave functions, but the underlying mechanism is the same.

That happens not only for the harmonic oscillator but also for electrons around atomic nuclei; their energy levels become discrete because of the behavior of the appropriate solutions to the Schrödinger equation, not because there is anything fundamentally discrete about space or time or energy or anything else. This is likewise the reason why quantum field theory, which is ultimately a theory of fields, looks like a theory of particles—those discrete quanta of energy are the solutions to the appropriately souped-up version of the Schrödinger equation.

The ultimate irony of quantum mechanics is that there's nothing fundamentally "quantum" about it. We see certain discrete things happen in the universe because that's how solutions to the Schrödinger equation work out.

TWO

MEASUREMENT

The Laplacian paradigm of classical mechanics says: you give me the state of the system, and the laws of physics will tell you how the system evolves from one moment to the next. You could be forgiven, having just read the previous chapter, for thinking that quantum mechanics follows the same basic pattern. Sure, the "state" is a wave function rather than a collection of positions and momenta for some particles, but you still specify it at some time, and the Schrödinger equation will help you evolve it to any other time.

That is not the whole story. Not even close. What makes quantum mechanics dramatically different from classical mechanics is that there are special moments when the central dynamical rule—the Schrödinger equation—does *not* seem to describe how the quantum state evolves. (Maybe it secretly does, but it certainly doesn't *seem* to.) Those moments are precisely when we observe, or measure, some property of the system.

This seems, upon a first encounter, outrageous. The mind struggles to comprehend how the mere act of looking at a physical system could play a profound role in the system's evolution. One reaches for classical analogies to help make sense of such a puzzling situation. Maybe it's like people who always make a funny face when you try to take their picture. Or maybe it's just that every observation involves a physical interaction of some sort; even when we look at the positions of planets in the sky or billiard balls on a table, our vision involves the exchange of photons, which carry a bit of momentum, which surely affects the trajectory of the object we're looking at.

These analogies are bad. What's going on in quantum measurement is much deeper and has no direct analogue in classical physics. Measurement is indeed a physical interaction, but in classical mechanics that interaction can be made arbitrarily tiny, so it disturbs the system as little as we like. In quantum mechanics even the slightest, most unobtrusive measurement can change the quantum state dramatically.

The linchpin of quantum mechanics seems to be this: what *exists* is not what is *seen*. They are related to each other, of course; it's not like anything goes. But when we measure a quantum system, we do not—and cannot, in general—simply observe what the quantum state was before the measurement. We observe some partial, incomplete aspect of the state, and in the process we change the state irrevocably.

This is why in quantum physics, as opposed to classical, "measurement" qualifies as one of the Biggest Ideas. As mentioned, there is little agreement on what's going on behind the scenes. Physicists and philosophers have labeled this **the measurement problem:** what is so special about measurements, when precisely do they occur, and what exactly happens when we make one? But we do have a good-enough-to-muddle-through picture of what experimental predictions quantum mechanics allows us to make, and that will be our focus here.

WAVE FUNCTION COLLAPSE

Schrödinger's introduction of the wave function and his equation for it was instantly popular among physicists striving to bring order to the chaotic mess of mid-1920s quantum theory. But there was an obvious looming problem: there are very clear circumstances under which electrons (and photons, for that matter) behave like particles, not like waves. How do we coax particle-like behavior out of the wave function?

Schrödinger himself had a hope: that when we start with a somewhat spread-out wave function and let it evolve according to his equation, the wave would naturally localize around some actual position, possibly moving as time passes. In other words, things could be fundamentally wave-like, but they might appear essentially particle-like because the wave function was zero almost everywhere, and nonzero just near some particular place.

That hope didn't work. The beauty and terror of a good equation is that there is no wiggle room. The equation determines what kinds of functions can solve it, and those solutions don't care about your hopes and dreams. It turns out that wave functions behave opposite of the way Schrödinger had wanted: if we start them relatively localized, they will spread out and become more diffuse over time. Wave functions like to be wave-like, not particle-like, at least as far as the Schrödinger equation is concerned.

Which is a problem, if we want to account for the world we see. Think of a radioactive atomic nucleus, one that decays into lighter particles. The Schrödinger equation is unambiguous about what the wave function of the newly created particles will look like: it will spread out in a diffuse cloud, moving in a roughly spherical shape away from the original nucleus. But that's not what we actually see. When a nucleus decays, we can take pictures of the emitted particles, and they describe definite trajectories, as we would expect from a moving particle. We see lines, not puffy clouds. This was one of the puzzles raised by Albert Einstein at the famous Fifth Solvay Conference in 1927, where the leading physicists of the day gathered in Brussels, Belgium, to codify the new rules of quantum theory.

When we teach quantum mechanics in modern textbooks, the explanation for this phenomenon hinges on the idea that the picture-taking—whether it is done by a cloud chamber in the 1920s, or a modern detector at the Large Hadron Collider outside Geneva—is an act of measurement. And measurement affects the wave function in a dramatic way.

Specifically, a quantum measurement is always the measurement of some specific quantity, the observable under consideration. An image of the motion of a newly created particle is a measurement of its position at each moment in time. The rule is this: whenever we measure an observable, whatever the wave function was before the measurement, it immediately **collapses** onto some definite value of the quantity being observed. The new post-collapse wave function then evolves according to the Schrödinger equation, until it is observed and collapses once again.

Here's how we solve Einstein's puzzle. The initial wave function of a particle created from a decaying nucleus is indeed emitted in a spherical shape. But as soon as it is detected by some observational process, it collapses to a particular location in space. That localized wave function also spreads out, but it mostly moves in the direction away from the original nucleus, because momentum is conserved. So when we observe it again soon thereafter, and then again a moment after that, the observations combine to describe something that looks just like the track (line) of a moving particle.

QUANTUM INDETERMINISM

How does the wave function know where to localize when we observe it? It doesn't. The position onto which the wave function collapses involves an irreducibly random element. But there is a definite **probability** for each possible measurement outcome, and that probability is determined by the wave function.

The probability is specified by the **Born rule**, first proposed by Max Born in 1926. The Born rule is pretty simple: the probability of measuring some position x is given by the square of the associated wave function:

$$P(x) = |\Psi(x)|^2. \tag{2.1}$$

We just have to be a little bit careful, because $\Psi(x)$ is a complex number, with both real and imaginary parts. That's why we put the vertical bars around the wave function in (2.1), to indicate that what we really have is the "modulus squared" of that complex number. If we break the wave function into real and imaginary parts, $\Psi = \Psi_R + i\Psi_I$, the modulus squared is given by the sum of the squares of those individual parts:

$$|\Psi|^2 = \Psi_R^2 + \Psi_I^2. \tag{2.2}$$

The modulus itself, $|\Psi| = \sqrt{|\Psi|^2}$, can be thought of as the "length" of the complex number, which we illustrated in the figure of the complex plane from the previous chapter. So where the wave function is big, in either its real part or its imaginary part, there is a relatively high probability of seeing the particle there if we measure its position. Where the wave function is small, there is a relatively low probability of seeing it there.

Let's think about why the modulus-squared of the wave function does the job that we expect a probability to do. Probabilities of events have two important properties: they cannot be negative, and when we add up the probability of every possible distinct outcome we have to get 1. (That's just how we express the idea that some unique thing is going to happen, even if we don't know which thing it will be.)

The non-negative property is pretty straightforward. The modulus-squared is the sum of the squares of two real numbers, which means that it cannot be negative. The wave function itself can't equal the probability, because the wave function can be negative (or imaginary). But the modulus-squared does the trick.

We also want the sum of all probabilities to add up to 1. But since we're measuring the position x of the particle, we don't have a discrete set of possible outcomes but rather a continuum. That's okay; we "add up" by doing an integral:

$$\int P(x)\,dx = \int \left|\Psi(x)\right|^2 dx = 1. \qquad (2.3)$$

We learned in *Space, Time, and Motion* that an integral is just the sum of the value of the thing being integrated at all locations. In English, this expression says "the total probability of seeing the particle at all possible locations equals one." The particle has to be seen somewhere, and it can't be observed to be at two places at once. We say that a wave function satisfying (2.3) is **normalized**. And if we start with a normalized wave function and let it evolve according to the Schrödinger equation, it stays normalized—the total probability of all possible outcomes remains 1.

Just to drive this home: according to textbook quantum mechanics, there are two different ways that wave functions can behave. When they are not being measured, they obey the Schrödinger equation. That behavior isn't too much different from what we encountered in classical mechanics: wave functions evolve smoothly, reversibly (information about the state is conserved), and deterministically. But at the moment of measurement, we throw Schrödinger out the window. The wave function collapses suddenly, irreversibly, and indeterministically, in accordance with the Born rule. There's nothing remotely like this in classical physics.

Schrödinger, who quite reasonably had hoped his equation would be the whole story, wasn't very pleased with Born's suggestion that it be used to calculate probabilities when wave functions collapse. "I don't like it," he grumbled, "and I'm sorry I ever had anything to do with it."

But there was some compensation (in addition to the Nobel Prize, everlasting fame, and so on). While it is true that the wave function of a particle freely moving through space will tend to become increasingly spread out, the rate of that spread depends on the mass of the particle. Lighter particles spread quickly, heavier ones more gradually. So when we work our way up to billiard balls or planets, the spreading is barely noticeable. For macroscopic objects, once they are localized in space they tend to behave pretty classically. Our good old Newtonian rules can be recovered in the proper circumstances, as a limiting case of the more general quantum framework.

WAVE-PARTICLE DUALITY

Nowhere is the weird duality of quantum evolution more evident than in the celebrated **double-slit experiment**.

As far back as 1801, Thomas Young performed a famous experiment to demonstrate the wave nature of light. This was before Maxwell's unification of electromagnetism, so it was still plausible to imagine that light was made of particles, and long before Planck and Einstein reintroduced something particle-like about light.

Young's experiment took advantage of a crucial property of waves: because they oscillate up and down, they can **interfere** with each other. If two waves come together, there will be places where they are both going up or both going down, in which cases they will join forces to go up or down even more ("constructive interference"). But at other places, one wave will be up and the other down, so they will cancel out ("destructive interference"). Particles can't destructively interfere like that; all you get by combining particles is more particles.

So Young set up an experiment that aimed a beam of light at two thin slits, very close to each other. And indeed, when he measured the light at a detecting plate on the other side of the slits, he observed an interference pattern of alternating light and dark bands. The part of the light wave that went through the left slit interfered with the part going through the right slit. This confirmed, according to the thinking of the day, that light was a wave.

The quantum version is a bit trickier. Imagine that we aim a single electron at two slits located close to each other. We know that the electron has a wave function, which will pass through the slits and presumably interfere with itself. But if we put a detector on the other side, it only observes a particle, which shows up as a dot with some definite location. So what does it even mean to say that the wave function interfered with itself?

To see, let's try this again with another electron and observe its dot. Keep doing this until the dots begin to accumulate to form an image. What we see is exactly the kind of interference pattern that we expect from wave-like behavior: bands where many dots have accumulated, separated by dark regions with almost no dots. If the electrons were simply classical particles, there would be no interference; instead, we would see two groups of dots, one corresponding to particles passing through each slit. But because each electron is a wave-like function right up until it gets detected on the screen, we see a pattern of light and dark bands that we expect from interference.

It gets better. The rule is that wave functions obey the Schrödinger equation, propagating as a wave, until they are observed, whence they collapse. So what if we observe "which slit the electron goes through"?

The interference pattern goes away. If we put a detector that measures the electron as it goes through the slits, the electron's wave function will collapse onto either "going through the left slit" or "going through the right slit." It will then continue on to the detector and leave a dot. But the locations of the accumulated dots will simply be those from the left slit, plus those from the right slit. There is no

MEASUREMENT

destructive interference, since the wave functions of each electron never went through both slits at once.

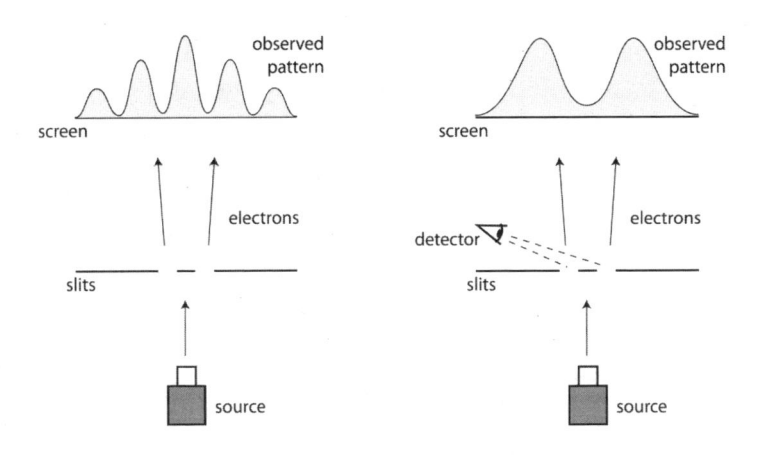

<figure>
Left: source → slits → electrons → screen showing multi-peak observed pattern (interference).
Right: source → slits → electrons with detector → screen showing two-peak observed pattern.
</figure>

Discussions of quantum mechanics often include a lot of mysterious discourse about "wave-particle duality." There is no reason for it to be confusing, at least at the level of what we see in the end. The electron acts like a wave when it is not being observed, and like a particle when it is, because observation causes the wave function to collapse to some particular location.

MEASUREMENT AND REALITY

Nevertheless ... there *is* something confusing and mysterious, if we want to dig a bit more deeply than "what we see in the end." Hopefully the above picture is enough to convince you that we know how to use quantum mechanics for almost all conceivable purposes. Set up a quantum system with some wave function, let it evolve according to the Schrödinger equation, then calculate the probability of different measurement outcomes using the Born rule. There's a bit of mathematical heavy lifting along the way, but the procedure itself is clear enough.

What's not clear is what is really going on. It's useful to distinguish between two issues about which there is little consensus among modern quantum physicists: the measurement problem, and the reality problem.

The **measurement problem** is, essentially, "What's really going on when we measure a quantum system?" Apparently it's something important, because we have to invent a whole new set of rules to describe what happens when we make a measurement. But we've been cagey about what actually qualifies as a measurement. Is it any interaction at all? Does it have to involve the exchange of information? Does it have to be carried out by a conscious agent? What if you look at a system but you're not really paying attention? How quick is the process of wave function collapse anyway?

It's kind of embarrassing to physicists, or at least it should be, that we can't agree on the answers to these questions. There are plausible answers, to be sure—the problem is that different theories answer the above questions in very different ways, and we don't agree on which theory is the right one. We'll talk a little more about the options in the next chapter, once we've thought a bit about quantum entanglement.

The **reality problem** is straightforward: "What is the correct quantum description of reality?" This isn't too much of a worry in classical physics. (Philosophers are able to worry about anything, but the worries associated with classical mechanics seem less pressing.) Reality is some set of particles and fields evolving in spacetime, following the laws of classical evolution. But in quantum theory, where we use the wave function to predict the probabilities of measurement outcomes, things are not so clear. Does the wave function itself represent reality? Or is it merely a useful calculational tool? Are there other physical quantities that we need in order to describe reality fully? Or is talk of "reality" somehow misleading, and we should just be happy to be able to predict what we are going to measure?

Again, there is no consensus among the experts. Some think that wave functions directly capture reality; others think that they need to be augmented by other variables that describe things like position and momentum; still others don't want to talk about reality at all, or they think that we only bring reality into existence by making measurements.

That's okay—scientists disagree about things all the time, and we make progress by thinking it over, collecting new data, and hashing it all out. But it's difficult to explain quantum mechanics in a satisfying way when we can't even agree on what the theory says.

Just as one example, consider again the double-slit experiment. When you think of an electron as a particle, it's tempting to reason along the lines of "the electron goes through either one slit or the other, we just don't know which." But if the wave function is the complete and final story, that reasoning is wrong: the electron's wave function goes through both slits at once. There is no such thing (under this view) as "which slit the electron goes through." Likewise, there would be no such thing as the position or momentum of the electron. Those are not quantities that exist, they are simply possible outcomes of observations. But observations don't give us an unvarnished view of the underlying reality.

At least, that's what we would say if we were "realist" about wave functions. Some people aren't. Wave functions are confusing, and we don't, as a community (or a species), agree on what they really are. Somehow we are perfectly able to use them to make predictions to amazing accuracy.

But we do need a way to *talk* about them, especially in a book like this. So although we will not be dwelling on foundational issues here, we will make the narrative choice of talking as if the wave function is a direct and complete representation of reality. This choice will come into play when we start talking about quantum fields and virtual particles, whose status also raises tricky philosophical questions. For our purposes, it is the wave function that represents reality, and all of our talk of "particles" and "fields" is a useful language for thinking about that reality in relatively intuitive terms. If your favorite formulation of quantum mechanics prefers different ways of talking, you are welcome to translate into whatever language you are comfortable with.

HILBERT SPACE

In classical mechanics, a central role was played by the concept of **phase space**—the set of states of a classical system. You could think of phase space as the set of all possible coordinates and momenta for whatever system you have in mind. An initial condition consists of a single point in phase space; that's all the information you need to solve the classical equations and find out what the system will do.

The analogous idea in quantum mechanics is **Hilbert space**—the space of all possible quantum states. Unlike classical states, when we have two quantum states we can add them together, or multiply them by (complex) numbers, and thereby get a different quantum state. So, for example, given two wave functions $\Psi_1(x)$ and $\Psi_2(x)$ and two complex numbers α and β, the quantity

$$\Psi_3(x) = \alpha \Psi_1(x) + \beta \Psi_2(x) \tag{2.4}$$

is also a wave function. (If all we do is multiply a single wave function by a number, the resulting state is physically equivalent to the original one.)

You can't add two classical particle locations together to get a third location. You might think you can, but that's because you're thinking of coordinates, and adding coordinates together depends dramatically on what coordinate system you are actually using. But there's no problem adding together wave functions. These states therefore satisfy the mathematical requirements for being considered "vectors" in some suitably abstract sense, so Hilbert space is an example of a **vector space**. A vector space is just a collection of objects (vectors) that you

MEASUREMENT

can add to each other or scale by multiplying by some number. So we've learned something about quantum states: mathematically, they are vectors, and the collection of all such vectors is Hilbert space.* What kind of Hilbert space it is will generally depend on the physical situation, just as phase space depends on what classical system we're thinking of.

It might seem like we are deploying words you've heard of before in an unfamiliar way. We've seen vectors: they're little arrows, with a length and a direction. In the vector spaces we've encountered before, we can draw axes and express the vector in terms of its component along each axis. How much of the vector is pointing in the x direction, how much pointing in the y direction, and so on. How do wave functions fit into that nicely intuitive picture?

The answer is subtle but not outlandish. The components of a wave function $\Psi(x)$ are simply the **amplitudes**—the (complex) values of the wave function—at each point x. Wait a minute, you say—there are an infinite number of points labeled by x. Indeed there are. The Hilbert space for a single particle is an *infinite-dimensional* vector space. There are also finite-dimensional examples of Hilbert spaces, but anytime we start with a classical system that could have an infinite number of locations (or other possible observational outcomes), the corresponding quantum Hilbert space is generally infinite-dimensional.

We're used to thinking of two-dimensional vector spaces, or three-dimensional, and maybe four-dimensional once we become sophisticated dwellers in spacetime. The leap to an infinite number of

---

* The term "Hilbert space" was introduced by Hungarian mathematical physicist John von Neumann, and named after David Hilbert, whom we met in *Space, Time, and Motion* as a colleague of Einstein's who derived the field equation for general relativity from the principle of least action. The complete definition of Hilbert space involves other mathematical conditions, including the ability to define a "dot product" between the vectors, but we won't need to worry about those details.

dimensions seems extreme and certainly is hard to visualize. The trick is that we don't try to draw an infinite-dimensional Hilbert space using an infinite set of orthogonal axes. We just think of one component for each value of spatial location x.

Paul Dirac, a brilliant British physicist who was one of the pioneers of quantum mechanics, came up with a nice notation for thinking about wave functions as vectors: we put the symbol for the vector between a vertical line and an angle bracket, like so: $|\Psi\rangle$. This is called a **ket**, which comes from the second half of the word "bracket."*

This notation helps us connect Hilbert space to the world of vectors as we've previously thought of them. Given an ordinary three-dimensional vector $\vec{v}$, we can express it in terms of components as

$$\vec{v} = v_x \vec{e}_x + v_y \vec{e}_y + v_z \vec{e}_z, \tag{2.5}$$

where $\vec{e}_x, \vec{e}_y, \vec{e}_z$ are basis vectors in the appropriate directions. We could translate this into Dirac notation as $|v\rangle = v_x|e_x\rangle + v_y|e_y\rangle + v_z|e_z\rangle$.

The analogous way to express the quantum state of a particle in one spatial dimension is

$$|\Psi\rangle = \int \Psi(x)|x\rangle \, dx. \tag{2.6}$$

Let's go through this expression slowly, as it is a paradigmatic example of how physics professors torture students by overloading similar-looking bits of notation with slightly different meanings. On the left we have $|\Psi\rangle$, which is the quantum state of our system. It has an existence in its own right, independent of any particular way of representing it, just as a three-dimensional vector like $\vec{v}$ is its own thing but

---

* Yes, there are also "bras," denoted $\langle\Psi|$. We can combine them to make a bracket of the form $\langle\Psi|\Psi\rangle = 1$, which is the slick Dirac-notation version of (2.3).

might have different components depending on the coordinate system we use.

On the right we have a particular expression for $|\Psi\rangle$ in terms of position x. Ordinary vectors can be expressed in terms of any axes you like, and quantum states will also be expressible in different ways. The kets $|x\rangle$ are the basis vectors, one for each value of the spatial location x, an infinite number of them overall. Just as we wrote the vector in (2.5) in terms of components v_i times basis vectors $\vec{e}_i$, summed over all three directions i, in (2.6) we have written the quantum state as a sum of amplitudes $\Psi(x)$ times basis vectors $|x\rangle$, summed over all values of x. (When we begin with an integral sign and end with dx, that means "sum the enclosed expression over all values of x.") Because the sum is over an infinite number of terms, we've written it as an integral, but the underlying concept is the same.

The fact that quantum states are vectors is a profound statement about the nature of reality, not just a bit of excess formalism. The basis vectors we've chosen, $\{|x\rangle\}$, correspond to something observable—the position of the particle. That's not a coincidence. Very generally, we can find a basis for Hilbert space in terms of states that have definite values for observable quantities. Here we're looking at position, but we could alternatively use momentum to define a different basis for Hilbert space.

If the quantum state of our particle just happened to coincide precisely with some basis vector $|x_*\rangle$, we would be guaranteed to observe the particle to be at the position x_* were we to measure it. In that case, we could say that the particle was in a state of definite position, or a **position eigenstate**, analogous to the energy eigenstates we considered in the context of the simple harmonic oscillator.

But most states are not of definite position; they are combinations of the form (2.6). So in those cases we say that the particle is in a **superposition** of many possible locations. ("Superposition" in the sense of "multiple positions, superposed," not "a really awesome position.")

That's a key feature of quantum mechanics, and one of the biggest motivations for people (unlike us) who don't like to think of the quantum state as representing reality. It's natural to associate "reality" with what we observe. When I look at particles, I see them in definite positions, so it's tempting to think that's what really exists. Quantum mechanics seems to be saying something profoundly different: that the state of the particle is in a superposition of many different possible measurement outcomes before we observe it.

QUBITS

Once you (or von Neumann) realize that quantum states can be thought of as kinds of vectors, it's natural to wonder whether the Hilbert space in which they live *has* to be infinite-dimensional. The answer is no: there are two-dimensional Hilbert spaces, three-dimensional ones, or those with dimensions of any larger integer. (Again, these "dimensions" are mathematical descriptions, nothing to do with directions in physical three-dimensional space.) And there is a natural context where these finite-dimensional Hilbert spaces are important: when we talk about the **spin** of a quantum particle.

In a famous experiment suggested by Otto Stern in 1921 and carried out by Walther Gerlach the following year, a beam of silver atoms was sent through a pinched magnetic field. A spinning particle can act like a little magnet, with a north pole and a south pole, and will be deflected in a magnetic field depending on how the axis of the magnet is oriented. We might expect the atoms to be deflected by various degrees, depending on the orientations of their spins. What we actually see in the Stern-Gerlach experiment is different: the atoms are either deflected up or deflected down. Nothing in between.

The reason why, as we now understand, is quantum mechanics. Each atom has a quantum state that depends not only on its position but also on its spin. When we send the atoms through the field and

MEASUREMENT

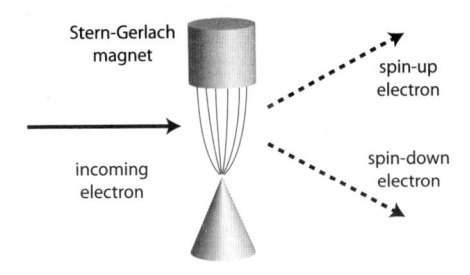

[Figure: Stern-Gerlach magnet showing an incoming electron being split into spin-up electron and spin-down electron paths]

see how they are deflected, we are measuring the amount of their spin along a particular axis. And in the case of silver atoms (as well as many elementary particles, including electrons, neutrinos, and quarks) there are only two possible measurement outcomes: "spin-up" or "spin-down." The angular momentum of a spin-up particle is $+\hbar/2$ and that of a spin-down particle is $-\hbar/2$, where $\hbar$ is our old friend the reduced Planck constant. When these are the only two possibilities, we say we have a **spin-½ particle** on our hands.

When we're describing the position of a particle in one dimension, there are an infinite number of possible positions x that we could observe, and the corresponding quantum states live in an infinite-dimensional Hilbert space, with basis vectors $\{|x\rangle\}$. When we're describing the spin of a spin-½ particle, there are two possible measurement outcomes, and correspondingly a two-dimensional Hilbert space to describe the quantum states. If we're measuring spin along the z-axis, for example, we can label the basis vectors as $|+z\rangle$ for spin-up and $|-z\rangle$ for spin-down, and write the state as $|\Psi\rangle = \alpha|+z\rangle + \beta|-z\rangle$, where α and β are complex numbers satisfying $|\alpha|^2 + |\beta|^2 = 1$. The basis vectors are "eigenstates of spin" (quantum states with a definite spin value, not a superposition of multiple possibilities) in the z direction.

We are also free to measure the spin along the x-axis, or the y-axis, or any other axis we like. We will, once again, have only two possible outcomes, spin-up or spin-down with respect to the axis we are using.

A spin-up state along the x-axis is $|+x\rangle$, spin-up along the y-axis is $|+y\rangle$, and so on.

Once we have specified the quantum state in some basis, writing it in a different basis does not require new information; it's just a repackaging of the existing information that specifies the state. In the case of spin, this can be seen because the basis vectors with respect to the x-axis or y-axis are just combinations of the z-axis basis vectors:

$$|+x\rangle = \frac{1}{\sqrt{2}}|+z\rangle + \frac{1}{\sqrt{2}}|-z\rangle,$$

$$|-x\rangle = \frac{1}{\sqrt{2}}|+z\rangle - \frac{1}{\sqrt{2}}|-z\rangle,$$

$$|+y\rangle = \frac{1}{\sqrt{2}}|+z\rangle + \frac{i}{\sqrt{2}}|-z\rangle,$$

$$|-y\rangle = \frac{1}{\sqrt{2}}|+z\rangle - \frac{i}{\sqrt{2}}|-z\rangle. \quad (2.7)$$

You can add and subtract combinations of these equations to solve for $|\pm z\rangle$ in terms of $|\pm x\rangle$ or $|\pm y\rangle$ as well. And then you can take the state written in any particular basis and swap in these relations to convert it to any other basis.

These unassuming equations are at the heart of the celebrated **Heisenberg uncertainty principle**, as we'll see in a moment. They remind us that it's the state that matters, not how we choose to represent it. The first line of (2.7), for example, indicates that a spin in the definite state $|+x\rangle$ can equally well be thought of as being in a certain superposition of $|+z\rangle$ and $|-z\rangle$ states. There is nothing like this in classical physics.

In the classical world, the basic unit of information we can contemplate is the **bit**: a number that has only two values, traditionally represented as 0 or 1. In the quantum world, the basic unit of information is naturally called a **qubit**, or "quantum bit." Rather than being 0 or 1, a qubit is represented by a superposition of two basis vectors in a two-dimensional Hilbert space, as in

$$|\Psi\rangle = \alpha|0\rangle + \beta|1\rangle. \qquad (2.8)$$

The complex coefficients α and β are called the amplitudes corresponding to $|0\rangle$ and $|1\rangle$, respectively. Likewise any particular value $\Psi(x)$ of the wave function for a single particle at location x is the amplitude for the particle to be at that position.

A bit of information is an abstract concept and can be represented in a variety of ways in a physical computer, from two positions of a switch to two different amounts of current in a circuit. Likewise, a qubit is an abstract state in a two-dimensional vector space, which can be represented by a variety of different physical systems. The spin of a spin-½ particle with respect to a given axis is one such representation. The physical interpretation of what we mean by the states $|0\rangle$ and $|1\rangle$ will depend on the technology we use to construct our qubits. Modern research in quantum computing has explored a variety of different ways of encoding qubits in physical systems.

MOMENTUM REVISITED

There is a curious imbalance in how we've been talking about wave functions. In classical mechanics, the state of a particle consists of both its position and its momentum, and those two quantities are completely independent of each other. If I tell you the position at just one moment in time, you have no idea what the momentum is. But we've presented the wave function as $\Psi(x)$—something that depends just on position. What happened to the momentum?

Don't worry, it's still there. But position and momentum are not independent variables as they were in classical mechanics. They are both observables, and we can calculate the probability of observing different values when we perform a measurement. And we can calculate the probability of either one from the single quantum state.

The apparent imbalance is entirely a matter of how we've chosen to introduce the wave function, as a complex number attached to every possible position we could measure the particle to be at. There is another way of expressing the quantum state that is just as valid: as a complex number for every possible *momentum* we could measure the particle to have. If momentum is denoted p, then we might write this alternative form of the wave function as $\tilde{\Psi}(p)$, where the tilde (~) over Ψ reminds us that it's a function of momentum, not position, even though it is representing the same physical quantum state. Then the probability of finding a certain value of momentum, were we to measure that, is just the Born rule as usual:

$$P(p) = |\tilde{\Psi}(p)|^2. \qquad (2.9)$$

The information contained in $\tilde{\Psi}(p)$ is precisely the same as that contained in the original wave function $\Psi(x)$. It is literally a change of basis, exactly as we can change the basis of a spin from $\{|+z\rangle, |-z\rangle\}$ to $\{|+x\rangle, |-x\rangle\}$. In the **momentum basis**, the equivalent of equation (2.6) expressing $|\Psi\rangle$ as a sum over position states is to express it as a sum over momentum states,

$$|\Psi\rangle = \int \tilde{\Psi}(p)|p\rangle \, dp, \qquad (2.10)$$

where $|p\rangle$ is a basis vector with definite momentum p (an eigenstate of momentum). Note that there is no tilde over Ψ in the ket on the left-hand side. That's because it's the *same* quantum state, not a new one. The only thing new is the way we've represented it as a sum over components, in this case the values of the momentum wave function $\tilde{\Psi}(p)$.

MEASUREMENT

The explicit formulas for carrying out the transformation from the position basis to the momentum basis and back are a bit involved, so we've relegated them to the appendix. The relevant mathematical trick is called the **Fourier transform**—a way of representing functions as a sum of sinusoidal waves rather than as explicit values at every location. The momentum-basis wave function $\tilde{\Psi}(p)$ is the Fourier transform of the position-basis wave function $\Psi(x)$. But we can certainly get some intuition for what's going on without losing ourselves in too many details.

A sound wave impinging on your eardrum can be thought of as a traveling oscillation in the density and pressure of the air around you. But when we listen to a piano or other musical instrument, our brains don't perceive every up-and-down variation of the air pressure. Instead, we hear musical notes—vibrations at specific frequencies. Depending on how musically adept we are, we might even be able to pick out the different notes that combine to create a chord. Similarly, when we tune a radio to a specific station, we're telling the radio to focus in on a particular wavelength of radio waves. There are many other waves out there; we're just ignoring them.

This is an example of a general idea that becomes useful whenever we are considering wave-like phenomena, whether in the form of sound waves or the quantum wave function. We can think of that wave as a function of space, oscillating up and down, *or* we can think of it as a certain frequency or combination of frequencies. So given any wave function $\Psi(x)$, we can express it as a weighted sum of waves of each possible wavelength. That's precisely what the Fourier transform accomplishes.

The connection with momentum is simple: a wave function that looks like a sine wave as a function of position is a state with a definite value of momentum—a momentum eigenstate. The relationship between the wavelength and the corresponding momentum is precisely de Broglie's equation (1.6):

$$\lambda = h/p. \tag{2.11}$$

So quantum states with *large* momentum are described by sinusoidal wave functions with *small* wavelength. This inverse relationship between size and momentum (and thus energy) will play a central role in quantum field theory.

THE UNCERTAINTY PRINCIPLE

Consider a wave function with definite momentum. As a function of x, it oscillates up and down with a fixed amplitude and precisely the same wavelength for all values of x. Such a wave function cannot possibly be normalized as in (2.3)—the integral would be infinite, since the function oscillates forever rather than fading away—but that's just one of the annoying mathematical details that we choose to gloss over to get to the important stuff. What we can say is that the probability of observing the particle to be at any particular position is spread out all over the place. When momentum is precisely known, we have no idea at all where the particle will be seen.

That's not just a quirk of these particular states, it's an example of a general rule, the Heisenberg uncertainty principle. We can think of states with definite momentum, but we will have no idea where the particle will be seen. On the other hand, a state with definite position corresponds to a superposition of all possible momenta, so we have no

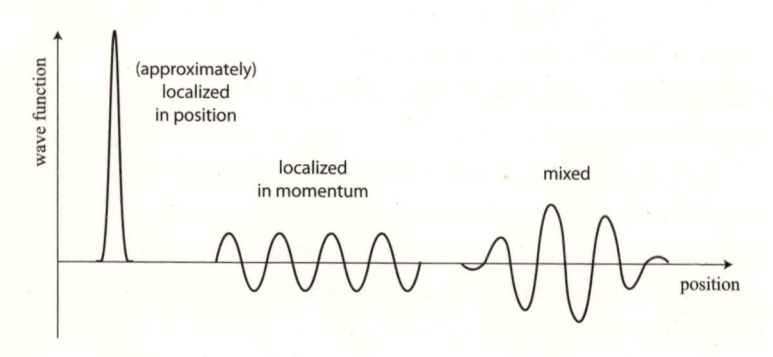

MEASUREMENT

idea what momentum we would observe. There are also "wave packet" states, with some spread in both x and p, where neither is determined precisely.

According to the uncertainty principle, in any conceivable quantum state, there is an irreducible minimum uncertainty associated with the combination of position and momentum, of the form:

$$\Delta x \, \Delta p \geq \frac{\hbar}{2}. \qquad (2.12)$$

Here Δx is the expected deviation in possible outcomes of a position measurement, and Δp is the expected deviation in possible outcomes of a momentum measurement. But the essence of the uncertainty principle isn't about measurements at all. It has implications for measurements—if we measure either position or momentum precisely, the wave function will collapse and we will have no idea what the other one would be immediately thereafter—but it's really a feature of quantum states even before we measure them.

The point is not that you inevitably bump into a quantum system while measuring it and therefore change it. It's that there do not exist any states in which both position and momentum are highly localized at the same time. This is hard to internalize if we remain stuck with classical intuition, thinking of position and momentum as things that really exist; it's easier to swallow if we think of them as sets of possible observational outcomes that we derive from an underlying quantum state.

Physicists sometimes express the opinion that the uncertainty principle is pretty trivial once you understand what is going on. It's just a change of basis, after all. We can write the quantum state as a sum over states of definite position, or as a sum over states of definite momenta, but a state that is definite in position is a superposition of momentum states, and vice versa. So of course it's impossible to have a state that is definite in both. It's just like the change of basis for spins, from z-axis to x-axis (or whatever) as in (2.7). A state of definite x-spin is a sum of up and down z-spins, and vice versa. A vector cannot point precisely along one axis and also precisely along another axis that is rotated at some angle with respect to the first.

That's only trivial once you've accepted the highly nontrivial part: that position and momentum (or x-spin and z-spin) do not have an independent existence. They, or at least the probability of observing them to have any particular value, are part of the single quantum state. It's the *state* that is physically relevant, not how we choose to express it in one basis or another. That's the profound thing.

Another thing you may hear is that quantum mechanics applies to the world of the very small. In fact, quantum mechanics applies to the whole world, big and small alike. But there is a classical approximation that describes the world well—extremely well—when objects are sufficiently large. That's not obvious from (2.12), which doesn't seem to refer to the size of the object at all. But it does refer to the momentum, which (for non-relativistic motion, much slower than the speed of light) is related to mass and velocity by $p = mv$. We generally take mass as a fixed quantity and can think of velocity as an observable just like momentum. Then we can write (2.12) as

$$\Delta x \, \Delta v \geq \frac{\hbar}{2m}. \qquad (2.13)$$

For a massive object, the uncertainty in the good old Newtonian parameters x and v can be extremely small when m is extremely large. That's why classical mechanics works so well in our ponderous, macroscopic world but breaks down when we consider tiny particles.

THREE

ENTANGLEMENT

The fact that wave functions seem to collapse when we observe them marks a profound difference between the classical and quantum ways of describing the world. It makes it hard for physicists to agree on whether the wave function represents reality or is merely a tool for making predictions. But this isn't the only thing about wave functions that makes them seem counterintuitive. It's not even the biggest thing. The biggest thing would be the fact that there aren't really wave *functions*, plural—as in, one wave function for every individual physical system. There is only one wave function for the entirety of everything, what we might (and do) call **the wave function of the universe**.

That's very different from classical mechanics. There, each part of the universe can be described separately, and combining systems is just a matter of including the appropriate description for each part.

Say we have one classical particle, A, moving in three dimensions. All the states it could possibly be in are captured by a six-dimensional phase space: three dimensions for A's position, and three dimensions for its momentum. That's the information we need to predict its future evolution. Now say we add another particle, B, to the system. This additional particle is described by its own six-dimensional phase space: three dimensions for B's position, and likewise for its momentum. The phase space for the combined system is just the sum of the two phase spaces, the one for A and the one for B—twelve dimensions in all. Keep adding more subsystems to our collection, and we just keep tacking on more dimensions to phase space. Each particle may interact with the others—that's a dynamical question, addressed by the equations of motion—but the *kinematic* question of how to describe their states at one moment in time is answered simply by "a point in a six-dimensional phase space for each particle."

Following this pattern, we might imagine that if we have one particle with a wave function $\Psi(x)$, we could add another particle by first putting a subscript on our original particle's wave function to make it $\Psi_1(x)$, then introducing a new wave function $\Psi_2(x)$ for the second particle. That's not how things work in general. The coordinate x is an observable—a possible measurement outcome—not a fact about the system with some definite value. So we should have an observable x_1 for where we see particle A, and a separate observable x_2 for where we see particle B. And the Schrödinger equation $\widehat{H}\Psi = i\hbar\,\partial\Psi/\partial t$ just has room for a single wave function in it, not a combination of two or more. So what we actually have is a single wave function that depends on both of the possible positions the particles can be observed in, $\Psi(x_1, x_2)$.

And that, in turn, means that what we might observe about one particle depends on what we might observe about the other, in a way that in classical mechanics it just doesn't. This is the most profound fact about quantum mechanics. Once you have really internalized this, many other things become clearer.

The usual way physicists express this feature of quantum theory is to say that two particles (or other subsystems of a bigger system) can be **entangled** with each other. The terminology is imperfect for a couple of reasons. First, when the human brain hears that two things are "entangled," it can't help but visualize some tangible connection between them, like a string or force field or something. Entanglement is subtler than that. In particular, particles don't exert a force on each other merely because they are entangled. And just by measuring one specific particle, there's no way of learning whether it's entangled with others.

Second, and more profoundly, thinking of a quantum system as a set of individual particles sharing a feature called "entanglement" is a relic of the classical picture, where the states of individual particles had some kind of independent reality. We can't help but think of the particles as what "really exists" and characterizing their quantum state as either being entangled or unentangled. But it's more helpful to think of the quantum state first and choosing to divide it into various particles as a later step that happens to be useful to humans, especially in the classical limit.

But the term "entanglement" is here to stay, and we'll continue to use it. The good news—maybe—is that the confusion engendered by the phenomenon of entanglement and the confusion engendered by quantum measurement might ultimately cancel out, because entanglement helps us understand measurement.

THE WAVE FUNCTION OF THE UNIVERSE

Let's see why entanglement is a natural consequence of the basic features of quantum mechanics.

The best way to get there is to think of the decay of a single particle into two particles. Think of the Higgs boson, postulated in the 1960s but only discovered in 2012 at the Large Hadron Collider experiment. The Higgs boson is exciting to particle physicists for reasons we'll discuss in later chapters, but for right now we invoke it just because it decays into other particles in a simple way: typically into two particles at a time, or really one particle and its antiparticle. So sometimes it will decay into an electron and its antiparticle the positron, other times into two photons (since photons are essentially their own antiparticles), sometimes into other things.

When we talk about the decay of the Higgs or some other unstable particle, we tend to think of it as a random process with a certain probability of happening at any time. What's really going on is that there is a quantum wave function representing a superposition of states of the form "the particle hasn't yet decayed" and "the particle has decayed."

$$|\Psi(t)\rangle = \alpha(t)|\text{not decayed}\rangle + \beta(t)|\text{decayed}\rangle. \qquad (3.1)$$

When we observe it, this wave function collapses and we see either the original single particle or the two particles into which it decays. According to the Schrödinger equation, the amplitude $\alpha(t)$ decreases with time and $\beta(t)$ increases. The probability that we see the original particle is $|\alpha|^2$, and the probability that we see the two decay products is $|\beta|^2 = 1 - |\alpha|^2$. So the longer we wait, the more likely it becomes to observe the particle as decayed.

Focus in on the last term in (3.1), the quantum state representing the two particles into which the original has decayed. Let's say one of them is an electron we'll label 1, and the other is a positron we'll label 2. If we run the experiment many times, observing the decays of many Higgs bosons, we'll see the created electrons and positrons moving in random directions away from their starting point. What that means—and what careful examination of the Schrödinger equation would confirm—is that the "decayed" wave function spreads out in a roughly spherical pattern away from the original particle.

So we can't predict what direction the electron will be observed in, nor what direction the positron will be observed in. But there is one further piece of information: momentum is conserved. If our original

ENTANGLEMENT

Higgs was just sitting there stationary with no momentum, the momentum of the observed electron must be equal in magnitude and opposite in direction to that of the observed positron, such that the two momenta add to zero.

But wait a minute. We can't predict the direction in which we'll observe the electron to be moving, nor that of the positron. Let's say we just put a detecting screen on one side of the original Higgs, and we happen to observe the electron hit it at some point. By comparing its position with the location of the original particle, we can figure out the direction in which the emitted electron was moving.

And that means we can, now, infer the direction that the positron is moving in, even if we haven't detected it. It must be moving oppositely from the electron. Somehow the wave function of the positron has collapsed from a spherical cloud moving in all directions to something localized with a particular momentum. How did that happen to the positron if we didn't even observe it directly?

The answer, as we've already revealed, is that the electron and positron aren't described by separate quantum states. There is a single state for the two-particle system. If x_1 is the location where we observe the electron and x_2 is the positron, in analogy with (2.6) we can write the combined wave function as $\Psi(x_1,x_2)$ and the state in Dirac notation as

$$|\Psi\rangle = \int \Psi(x_1,x_2) | x_1,x_2 \rangle \, dx_1 \, dx_2, \qquad (3.2)$$

where $|x_1,x_2\rangle$ is a basis state where each particle has a definite position.

$\Psi(x_1,x_2)$ tells us the probability that we simultaneously observe the electron at x_1 and the positron at x_2, namely $P(x_1,x_2) = |\Psi(x_1,x_2)|^2$. We can certainly think of situations in which two particles have nothing to do with each other, and their state is unentangled. But in other situations, such as when both particles arise from the decay of a single one, the probabilities are very much related. That's entanglement.

This pattern generalizes. No matter how big our system is, there is only one wave function. Ultimately there is just one quantum state for everything, the wave function of the universe. It is often possible to subdivide it because different pieces happen to be unentangled, but that's not the generic situation.

Two things to note about this entanglement story. First, it makes clear that wave functions aren't fields, in the way we have talked about fields in classical mechanics. The whole point of a field is that it takes on a unique value at every point in space. For a single particle, when the wave function looks like $\Psi(x)$, that's basically what happens, so you'd be forgiven for thinking of the wave function as a kind of field. But as soon as we have two particles, $\Psi(x_1, x_2)$ does not assign a number to every point in space: it assigns a number to every *pair* of points. And for three particles, the wave function $\Psi(x_1, x_2, x_3)$ assigns a number to every triplet of points, and so on. Wave functions are not functions of space; they are functions of the *configuration space* for whatever system we're thinking of. (Or of the momentum space, if we want to represent them that way.) That means that wave functions are not fields in the usual sense. This is what allows for the possibility of entanglement.

The other thing to note is that (3.1), representing the superposition of a single undecayed particle and its two decay products, seems a little fishy if we really think about it. The Hilbert space for a single particle consists of complex-valued functions of one variable, $\Psi(x)$, while the two-particle Hilbert space is complex-valued functions of two variables, $\Psi(x_1, x_2)$. Is it legitimate to just stick them together into a single Hilbert space?

The answer is that it can be made legitimate, but in order to describe transitions between different numbers of particles we have to turn to quantum field theory. We'll get there soon.

SPOOKY ACTION AT A DISTANCE

Entanglement was first emphasized as a feature of quantum mechanics in a famous 1935 paper by Albert Einstein, Boris Podolsky, and Nathan Rosen, which has subsequently come to be known simply as EPR. It arose out of Einstein's conviction that quantum mechanics could not be the correct final story when it came to the workings of the universe. The point of the paper was not that quantum mechanics was wrong but that it was somehow incomplete. It's right there in the title: "Can Quantum-Mechanical Description of Physical Reality Be Considered Complete?" (The absence of the word "the" reflects the house style of the journal *Physical Review*, not the fact that Podolsky was from Russia. I think.)

EPR considered a situation much like our decaying particle, but it's easier to make the basic point using entangled spins rather than particle positions. A Higgs boson has zero total spin, while the electron and positron are both spin-½ particles. By conservation of angular momentum, the electron and positron need to have opposite spins, since they need to add up to zero.

Entanglement allows us to consider a state where the spin of either particle is completely uncertain, but the two of them are constrained to be opposite. Let's consider measuring spin along the z-axis, and we'll write a spin-up state as $|\uparrow\rangle$ and a spin-down state as $|\downarrow\rangle$. For the two-particle electron/positron system, we can define four basis states of definite spin for both, with the first arrow representing the electron and the second one for the positron:

$$|\uparrow\uparrow\rangle, |\uparrow\downarrow\rangle, |\downarrow\uparrow\rangle, |\downarrow\downarrow\rangle.$$

A state like $|\uparrow\downarrow\rangle$ means that we are guaranteed to find the electron spin-up and the positron spin-down when we measure them along the z-axis. It has zero total angular momentum, but the two spins are *not* entangled: in that state we wouldn't learn anything new about the electron by measuring the positron, or vice versa. By taking the correct superposition, however, we can construct a state where neither spin is known ahead of time, but the total angular momentum is guaranteed to be zero:

$$|\Psi\rangle = \frac{1}{\sqrt{2}}(|\uparrow\downarrow\rangle - |\downarrow\uparrow\rangle). \quad (3.3)$$

This is the kind of state the spins will be in when the particles are emitted by a decaying Higgs. If we measure the spin of the electron, the probability that we will see the $|\uparrow\downarrow\rangle$ state is given by its amplitude squared, $(1/\sqrt{2})^2 = 1/2$. In that case, the electron is spin-up. The state as a whole will have collapsed onto the part where the electron is spin-up, namely $|\uparrow\downarrow\rangle$. So now we know for sure that the positron would be spin-down if we also measure it along the z-axis. And likewise if we see the electron to be spin-down, we know the positron will be spin-up. Measuring part of the wave function has dragged the rest of it along for the ride, due to entanglement.

There is a popular myth to the effect that Einstein, as great a physicist as he was, had become an old fuddy-duddy by the time quantum mechanics blossomed, and he was prevented from fully appreciating the theory by various philosophical prejudices. Nothing could be further from the truth. As we have seen, Einstein helped launch the quantum revolution, and in 1935 he understood quantum mechanics as well as anyone. He just didn't think it was the final answer. And while it's true that he once complained that God doesn't play dice with the universe, the random nature of quantum predictions wasn't his primary stumbling block. He was much more interested in *realism*—that there really is a physical world out there, independent of our observations—and *locality*—that there is such a thing as "spacetime," and what happens in the world happens at individual locations in spacetime.

The EPR thought experiment brought realism and locality into conflict, involving what Einstein memorably dubbed **spooky action at a distance**. After the Higgs decays, the electron and positron move apart from each other. We could capture them without measuring their spins, keep one nearby, and send the other one on a rocket ship traveling light-years away. Let's say Alice is here in a laboratory on Earth with the electron, and Bob is orbiting Alpha Centauri with the positron. Alice finally measures the electron's spin, and let's imagine she registers it as spin-up. Before that measurement, we had no idea what Bob would see for the positron. After, Alice knows for sure he will see spin-down. The state of the positron has changed instantly, no matter how far away it is. Einstein, of all people, knew perfectly well that no actual information can travel between the particles faster than the speed of light, by the rules of relativity. The word "instantly" isn't even well defined at widely separated locations. So how does the positron "know" how to respond to the measurement of the faraway electron?

This turns out to be more a violation of the spirit of relativity than the actual letter of the law. Sure, Alice knows what the positron spin will be as soon as she measures the electron. But Bob doesn't learn anything he didn't already know, since he's too far away to see what Alice measured. As far as he is concerned, there is still a ½ chance of seeing spin-up or spin-down when he measures the positron. Alice could send him a message telling him what her measurement result was, but that message would have to travel by ordinary, slower-than-light means. Despite the active imaginations of people who know a little bit about quantum mechanics but not quite enough, entanglement does not allow us to send information faster than light.

THE EPR PUZZLE

There's a temptation to imagine that maybe entanglement is not as spooky as it's made out to be. Imagine you have two marbles, one red and one blue, and you place each of them in identical-looking boxes.

A friend mixes up the two boxes so you don't know which is which, hands you one, and then takes the other far away. You open your box and see the red marble. Now you instantly know the other one is blue, even though it's far away and you didn't know that before. What's so spooky about that?

That is not what is going on in quantum mechanics. Entanglement is more than just a simple correlation between things that already exist but are unknown to you. We know this in part because the wave function acts like a physically real thing—for example, by interfering in the double-slit experiment. But more directly, the kinds of correlations entailed by quantum entanglement go well beyond simple classical relationships.

EPR appreciated this, so they tried to turn their discomfort into something more explicitly paradoxical. They posit that, when we can say with absolute certainty what a measurement outcome will be, there must be an "element of reality" corresponding to that property of the system. By this definition, after Alice measures the electron to have spin-up, there is an element of reality that says "Bob's positron will be spin-down." And according to EPR, that element must have been there all along, since nothing can travel faster than light.

But Alice didn't have to measure her electron's spin along the z-axis; she could equally well have measured along the x-axis, obtaining possible results "spin-right" or "spin-left" rather than up and down. Let's label the corresponding states $|\rightarrow\rangle$ and $|\leftarrow\rangle$ (which we called $|+x\rangle$ and $|-x\rangle$ in the last chapter). Comparing the relationship of z-spins to x-spins from (2.7) to the entangled state (3.3), it's pretty easy to verify that the state can also be written as

$$|\Psi\rangle = \frac{1}{\sqrt{2}}(|\leftarrow\rightarrow\rangle - |\rightarrow\leftarrow\rangle). \qquad (3.4)$$

This is the same quantum state as (3.3), just written in a different basis. The two spins are still entangled, and they are still oppositely aligned: if Alice measures spin-left, Bob will inevitably see spin-right, and vice versa. That makes sense, given that the angular momentum is supposed to add up to zero.

EPR claimed that it's easy to imagine a situation where Alice would have measured spin-up had she measured the z-spin, and she would have measured spin-left had she measured the x-spin. Therefore, by their lights, there must be separate elements of reality fixing both the z-spin and the x-spin of Bob's particle, and (because of relativity) those elements must have been there all along. But that violates the uncertainty principle. You can't have a state that is both in a definite value of z-spin and x-spin at the same time, according to the traditional rules of quantum mechanics. Therefore, EPR concluded, quantum mechanics itself must be incomplete.

Faced with the EPR argument, the standard response on the part of a modern physicist is simply to accept that there is some kind of influence that originates from Alice's measurement event and spreads instantaneously (whatever that means) to Bob's, even if the two events are spacelike separated (outside each other's light cones). As mentioned, there is no way to take advantage of this spooky action at a distance to send signals or information, even if the formalism we use seems to invoke faster-than-light influences. Part of the standard response is to not ask too many uncomfortable questions about this situation and change the topic to other matters.

MEASUREMENT AND ENTANGLEMENT

At the beginning of this chapter we teased the possibility that entanglement might help us make sense of the mystery of quantum measurement. Let's see how the two might be related.

To get there, let's be more explicit about what happens during a measurement. We've said that a quantum system can be in a superposition, but when a quantity is measured, we only ever see that quantity having some definite value. To perform such a measurement, we need some kind of apparatus. In the Stern-Gerlach experiment that measures spin, for example, we input a particle in a superposition of spin-up and spin-down, and the apparatus outputs a definite answer, either "up" or "down." We might be even more concrete and imagine a dial with a big pointer on it, and labels *UP*, *Ready*, and *DOWN*. Before the measurement, the pointer aims at *Ready*, indicating that we haven't started yet. After the measurement, it points toward either *UP* or *DOWN*, depending on the outcome.

Our apparatus is presumably made of atoms and particles and so on, so it should obey the rules of quantum mechanics just like a single spinning particle should. The full quantum description of a macroscopic laboratory apparatus may be intimidatingly complicated, but we won't lose anything important for our present purposes by abstracting it down to just three basis states: $|UP\rangle$, $|Ready\rangle$, and $|DOWN\rangle$, each associated with the corresponding definite pointer position. We are also free to imagine superpositions of these states, of course, but we never run across them in real laboratories. The pointers with which we are familiar always seem to be pointing in definite directions.

We can start with a particle in some superposition, $|\Psi\rangle_{\text{particle}} = \alpha|\uparrow\rangle + \beta|\downarrow\rangle$, and then let it be measured by an apparatus in the initial state $|\Psi\rangle_{\text{apparatus}} = |Ready\rangle$. The full initial state is just the product of these two, which we can write as

$$|\Psi\rangle_{\text{initial}} = (\alpha|\uparrow\rangle + \beta|\downarrow\rangle)|Ready\rangle. \qquad (3.5)$$

We could consider initial situations in which the particle and the apparatus are already entangled, but we're choosing to consider an unentangled one.

When the measuring process is complete, things have collapsed to either have the particle in a completely $|\uparrow\rangle$ state and the apparatus in

ENTANGLEMENT

the $|UP\rangle$ state, or the particle in a completely $|\downarrow\rangle$ state and the apparatus in the $|DOWN\rangle$ state.

$$|\Psi\rangle_{\text{final}} = |\uparrow\rangle|UP\rangle, \quad \text{probability } |\alpha|^2,$$

$$|\Psi\rangle_{\text{final}} = |\downarrow\rangle|DOWN\rangle, \quad \text{probability } |\beta|^2. \quad (3.6)$$

All of this collapse business, as we know, is in flagrant violation of the Schrödinger equation, and it's not clear how, why, or precisely when the collapse happens.

So let's ask: What would happen if we *didn't* violate the Schrödinger equation? You might suppose that's an impossibly hard question, since the system of atoms and molecules that make up a macroscopic apparatus is too complex to characterize precisely. But we have an enormous simplification in front of us, from the basic fact that we want our measurement to be accurate. That means that if the initial particle state was $|\Psi\rangle_{particle} = |\uparrow\rangle$ (pure spin-up, no admixture of spin-down), then the final apparatus state had better be pure $|UP\rangle$ with 100 percent probability, and likewise for spin-down. The only way that can happen is if the more general initial state (3.5) evolves into

$$|\Psi\rangle_{\text{final}} = \alpha|\uparrow\rangle|UP\rangle + \beta|\downarrow\rangle|DOWN\rangle. \quad (3.7)$$

In other words, what happens to a good measuring device is that it becomes entangled with the system being measured, in just such a way as to associate proper readouts with the corresponding system states.

That's not what real-world scientists actually see, though. What we see is indisputably particles and pointers in some definite state, as in (3.6). So the measurement problem is not so much solved as sharpened: What disrupts ordinary Schrödinger evolution so that we seemingly end up with (3.6) instead of (3.7)?

DECOHERENCE

There is one dangling thread. We made a big deal about the fact that there is only one wave function for the whole universe, but we have continually analyzed situations containing just a few moving parts, ignoring everything else in the world. That's okay when we're thinking about microscopic particles, which can be separated from the rest of the world for some reasonable period of time. But when a macroscopic object like a measuring apparatus is involved, this spherical-cow maneuver breaks down. Like it or not, pieces of the rest of the world—molecules in the air, ambient photons in the room—are going to keep bumping into the apparatus. Does that matter?

It does. Let's reanalyze the measurement of a single spin, but now let's include everything else in the universe, wrapped up into a single system we label the **environment**. That includes air molecules, photons, but also the experimenter and the table and everything else. Think of the environment as all of those things that we don't explicitly keep track of during the experiment. Once again, the state of the environment will be unimaginably complex, but we care about only a small number of features.

Think about how our measuring apparatus interacts with its environment. The experimenter, if they are careful, will mostly leave it alone during the measurement, so nothing interesting happens. But the photons in the room will be important, since we need them to illuminate the dial. And—this is the crucial part—those photons will interact differently with the apparatus depending on what the apparatus is doing. We have imagined a pointer that could be pointing to the different readouts: *UP, Ready*, and *DOWN*. Let's say the pointer is black (so it absorbs photons), while the background of the dial is

ENTANGLEMENT

white (so it reflects). Then it's easy to imagine how a certain photon will hit the pointer if it's pointing toward *UP* and be absorbed, but the same photon will be reflected by the dial if the pointer is in some different direction.

This, in turn, means that the bath of photons in which the apparatus is necessarily immersed will not only interact with the apparatus; it will become entangled with it, almost immediately.* Let's say the environment starts in an initial state $|e_0\rangle$, then evolves to $|e_\uparrow\rangle$, or $|e_\downarrow\rangle$, depending on whether its photons are bumping into an arrow pointing at *UP* or *DOWN*, respectively. Then the combined spin/apparatus/environment state immediately post-measurement will look something like

$$|\Psi\rangle_{\text{measurement}} = (\alpha|\uparrow\rangle|UP\rangle + \beta|\downarrow\rangle|DOWN\rangle)|e_0\rangle, \qquad (3.8)$$

but photons move at the speed of light, so this will evolve extremely quickly into

$$|\Psi\rangle_{\text{decohered}} = \alpha|\uparrow\rangle|UP\rangle|e_\uparrow\rangle + \beta|\downarrow\rangle|DOWN\rangle|e_\downarrow\rangle. \qquad (3.9)$$

So it's not only the measurement apparatus that becomes entangled with the system being measured. Because the apparatus is a macroscopic object that cannot help but interact with the bath of photons around it, the environment becomes entangled too. This process is known as **decoherence**—when a quantum system in a superposition becomes entangled with its environment. For large objects that keep

---

* You can interact without getting entangled; entanglement occurs only when different states in one system become associated with different states in another. The state (3.7) is entangled, while (3.5) is not.

bumping into the rest of the world, decoherence is rapid and nearly unavoidable.

Why should we care if macroscopic systems are continually becoming entangled with their environment? Think back to the double-slit experiment. When we didn't observe the particle as it passed through the slits, we would see interference patterns emerge at the detector screen on the other side. That interference arises because there are two contributions—one for each slit—to precisely the same final amplitude for a particle to be observed at some particular position. When the particle is "measured" as it goes through the slits, it is actually just becoming entangled with some external measuring device. That means when it reaches the detector screen, the contributions from the two slits are not directly comparable. They are entangled with different states in the external world, so they can neither cancel each other out nor add to each other. Decoherence destroys interference, which is one of the most fundamental aspects of quantum behavior. And decoherence is irreversible—the environment has such a large, chaotic collection of moving parts that once it becomes entangled with a system, the chance that it will ever become disentangled is negligibly small.

This suggests an answer to what a quantum measurement actually is: it's when the system decoheres. Measurement, as it appears in quantum mechanics, doesn't have anything to do with consciousness or perception. It's just what happens when a quantum system in a superposition becomes entangled with its environment.

FOUNDATIONS

That doesn't mean that decoherence "solves the measurement problem." It explains why the interference pattern goes away in the double-slit experiment and therefore suggests a nice physics-based understanding of what we mean by a "measurement." But at the end of the day, simply following the Schrödinger equation leaves us with the entangled state (3.9), while what we apparently see is one of the collapsed states in (3.6). We still need to somehow explain why we see the latter rather than the former.

This is the topic of **foundations of quantum mechanics**—what is the complete and unambiguous theory that explains the correct predictions of the somewhat ad hoc formalism of vaguely defined "measurements" and sudden "collapses"? We don't yet agree on what the answer is. Here at the Biggest Ideas in the Universe our focus is on things that we do agree on, so we'll just offer a quick overview of some of the leading ideas.

Here's one: maybe the wave function represents reality, and it always obeys the Schrödinger equation. That would certainly be nice, from the perspective of aesthetics and simplicity. The problem is that we end up with (3.9)—a quantum state that is in a superposition of different measurement outcomes. And no experimenter has ever felt like they were in a superposition of different measurement outcomes. Real experiments seem to have definite results.

One way to solve this problem was suggested by Hugh Everett III when he was a graduate student in the 1950s. Everett proposed that the Schrödinger equation is fine, what's mistaken is our identification of "the experimenter" in a state like (3.9).

Remember that once decoherence happens, in practice the parts of the state that are entangled with different environment states will never interfere with each other again. In that case, says Everett, the right thing to do is to treat them as completely independent copies of the universe, called **branches**. There is not one experimenter who is in a superposition; there is one experimenter in the $|UP\rangle$ branch of the wave function, and another in the $|DOWN\rangle$ branch, both of whom experience definite measurement outcomes. This has come to be known as the **Many-Worlds interpretation** of quantum mechanics (although it's an honest physical theory in its own right, not an "interpretation" of anything).

Many-Worlds is simple and compelling but raises people's eyebrows for obvious reasons. A different approach is to posit that the wave function represents *part* of reality, but not the whole thing—that there are other variables needed to fully describe the physical configuration of the world. In the case of a collection of particles, those variables might simply be the actual positions of the particles. This approach was pioneered by Louis de Broglie himself, and later revived and popularized by David Bohm, so it is often known as **de Broglie–Bohm theory** or simply **Bohmian mechanics**. It is a kind of **hidden-variable theory,** because we need extra variables to tell us where the particles are. It's also known as a **pilot wave** theory, because the role of the wave function is to "pilot" the particles. They tend to congregate where the wave function is large and scatter away from where it is small.

Hidden-variable theories have a curious history. In 1935, John von Neumann presented a supposed proof that they could never work. Grete Hermann pointed out that von Neumann had made unnecessary assumptions, but she was largely ignored, and physicists assumed that the possibility had been eliminated. In the 1950s, Einstein pointed out the limitations in von Neumann's argument to Bohm, who set about inventing a viable model and came up with one that was similar to de Broglie's from years before. This was also largely ignored, but it caught the attention of Irish physicist John Bell, who was a theorist at CERN, the particle-physics laboratory in Geneva, Switzerland. Bell noticed an important feature of Bohm's theory: it was **nonlocal**, because the way in which the wave function pushes the particles around depends on the configuration of all the particles, not just the one being pushed. Bell wondered whether this feature was necessary, then proved that it is: **Bell's theorem** establishes that it's impossible to reproduce the predictions of quantum mechanics in a hidden-variable theory that is entirely local.* The Nobel Prize in Physics was given in 2022 to Alain Aspect, John Clauser, and Anton Zeilinger for experimentally demonstrating that entangled particles in the kind of setup that Bell envisioned really did obey the predictions of standard quantum theory. Unfortunately there was a mistake in the Nobel press release, claiming that Bell's theorem eliminated the possibility of hidden-variable models. It really eliminates only the possibility of *local* hidden-variable models. Bell himself would have been mortified, as he was a big fan of Bohm's nonlocal theory.

Yet another viable approach to quantum measurement is simply to accept that wave functions really do collapse, even if this violates the Schrödinger equation. There are a number of different strategies along these lines, depending on whether the collapses are spontaneous and random or whether they are triggered by some feature of the quantum state. The good news is that these **objective collapse models** are experimentally testable. The bad news, at least thus far, is that there is no experimental evidence in their favor.

Finally, some physicists prefer to hark back to the original Copenhagen philosophy of Bohr and Heisenberg. The idea here is to stop worrying about what is "real" and shift focus onto what we actually see: measurement outcomes. In **epistemic approaches** to quantum foundations, wave functions are merely tools for agents to calculate the probabilities of measurement outcomes. Then there is nothing spooky about their collapse; it is simply the usual process of updating our beliefs when new evidence comes in.

The variety of options on the table reflects the inability of the

---

* Like any mathematical theorem, Bell's comes with assumptions. One is that experiments have definite outcomes; this is violated in Many-Worlds, where different outcomes happen in different branches. Another is that we are able to measure any quantity we might want to measure; this is violated in "superdeterministic" theories, where the initial conditions of the universe guarantee that certain properties will never be measured.

physics community to agree on the foundations of quantum mechanics. Physicists are extremely good at using a theory to make predictions, less good about establishing what the theory actually says.

For our purposes in this book, we therefore have a situation: How are we going to talk about wave functions and reality, given that professional physicists don't agree on the right way to do so? We could, at every relevant juncture, indicate a number of different ways to express what is fundamentally going on, depending on one's attitude toward the foundations of quantum mechanics. But that would be tedious. Instead, we're going to pick one attitude and try to be consistent about it: namely, that the wave function directly and uniquely represents reality. And, as we'll see, that wave function is built from quantum fields pervading spacetime. In this picture, electrons are not really point-like particles, they are quantized vibrations in the electron field; atoms are not mostly empty space, they are defined by a certain wave function profile. When we get to interactions and Feynman diagrams, we will mention that "virtual particles" are even less literally real; they are a convenient device for calculating the probability of certain quantum processes. But they are a very convenient device, and it's completely okay to talk about them, as long as we remember what's really going on under the hood.

Keep in mind that alternative perspectives exist. You are welcome to translate the language used here into whatever would be suggested by your favorite approach to quantum foundations. There is certainly room for future generations to clarify the situation in important ways.

FOUR

FIELDS

Like classical mechanics, quantum mechanics is not a specific physical theory by itself. It's a framework, within which we can build specific models of different kinds of systems. You have a classical theory of the simple harmonic oscillator, and also a quantum theory of the simple harmonic oscillator. Very often in physics we have a kind of physical system in mind, then think about how to construct a quantum theory of that system.

In classical mechanics we distinguish between particles, which have particular positions in space, and fields, which take on values at every point in space. We can do the same thing in quantum mechanics. In the last few chapters we've been considering quantum theories of particles. The quantum mechanical theory of fields is quite sensibly known as **quantum field theory**, or **QFT** for short. Some folks speak as if there was first "quantum mechanics," and then it was superseded by "quantum field theory." That's not accurate; QFT is part of quantum mechanics, just applied to fields rather than particles.

As of this writing, quantum field theory is the single best way we have of describing the universe at its deepest known level. We don't even need to consider particles and fields separately; when we think carefully about quantum fields, the particles will just pop right out at us. The quintessential example of "quanta" appearing from a fundamentally smooth picture of the world.

Consider a simple field, which we will denote by ϕ (Greek letter phi). Being a field, it will have a **value** at each point x in space, $\phi(x)$. That value might be zero, but it's still a value. (For reasons of notational simplicity, we're going to write everything as if there is just one dimension of space, x, but all the ideas work equally well in three dimensions.) Even if space is as empty as space can be, the fields are still there. This is in contrast with particles, which have a location in space at each moment in time, but no kind of existence at the other points where they are not located.

It is often useful to consider an entire **field configuration**, or "profile," which specifies the actual values the field has at every point in space. We'll denote field configurations using braces, $\{\phi(x)\}$. But sometimes we get lazy and skip the braces (and other references don't use them at all), so stay vigilant. As the field evolves through time in accordance with an appropriate equation of motion, it will attain values $\phi(x,t)$ at each point in spacetime. Mathematically, we might say that a field is defined by a map from space to some kind of values. For the moment we'll stick with the simplest case, a **scalar field**, where values are simply numbers at each spatial point. Later we'll consider vector fields and more exotic possibilities.

As we'll see shortly, the best way to wrap our heads around quantum fields is not actually to think of them as values at each point in space but to express a field configuration as a sum of waves with different wavelengths (via the Fourier transform). Each term in that sum is a "mode" of a definite wavelength, much like a musical tone of definite pitch.

And then the miracle occurs. Each mode of a quantum field behaves like a simple harmonic oscillator, including the quantized energy

FIELDS

levels we previously uncovered. Those energy levels are interpreted as the number of *particles* we would observe: a mode in its first excited state represents one particle, its second excited state represents two particles, and so on. It's a pretty magical quantum connection between fields and particles, and it will be worth getting our hands a bit dirty in the mathematical weeds to see it happen.

It's tempting to wonder what fields "really are." Like, what are they made of? There's no satisfying answer to that question. In the context of quantum field theory, fields aren't "made of" anything—they are the things that everything else is made of. We shouldn't be surprised by that. If we dig sufficiently deeply into what things are made of, we're bound to bottom out with "here's the bare stuff of reality, which is not made of anything else."

According to our current best understanding, quantum fields are the bare stuff of reality. It's completely conceivable that in the future our understanding will improve, and we'll come to appreciate that fields aren't fundamental after all. There are even some hints of this from modern attempts to understand quantum gravity. But that's still quite speculative, and for Biggest Ideas purposes it makes sense to stick with the established paradigm of quantum fields.

THE ENERGY OF A FIELD

Most often, when physicists construct a quantum theory, they start with a classical theory and quantize it. That's true for fields just as much as it was for particles. So let's think about a classical field and ask what kind of energy such a field might carry. That will lead us to an appropriate quantum Hamiltonian, which powers the Schrödinger equation.

For a single particle we considered two types of energies: the kinetic energy $\frac{1}{2} mv^2$, and the potential energy $V(x)$, where m is the particle's mass, v is its velocity, and $V(x)$ is the potential-energy function. The kinetic energy is associated with motion, and the potential energy just depends on the position of the particle and external factors that are influencing it. When it comes to fields, there's no such thing as "the position," but there is the value $\phi(x,t)$ that the field has at each event (x,t), as well as the values at other points in space and time. We need to construct something energy-like out of that information.

A major constraint on how we can do this comes from the principle of **locality**. We talked about locality a bit in the context of EPR and quantum measurement, but now we're just thinking about the classical behavior of fields. That behavior is entirely local, nothing spooky about it.

For a field, locality means that how it evolves at any one point in spacetime only depends on the value of that field and other fields at that same point, as well as the immediate neighborhood of that point. "Immediate neighborhood" is a way of saying that we can appeal to the *derivatives* of the field in space or time (since they reflect what the field is doing infinitesimally nearby), but the behavior of the field at (x,t) doesn't depend directly on anything happening at any other point a finite distance away. What this amounts to in practice is that the behavior of a field ϕ at location (x,t) can be affected by the field value itself $\phi(x,t)$, as well as its partial derivatives with respect to space and time:

$$\frac{\partial \phi}{\partial x}, \quad \frac{\partial \phi}{\partial t}.$$

Remember that the partial derivative just says "take the rate of change with respect to this particular variable, keeping all of the other variables fixed." So at each point (x,t) in space and time, the field will have spatial derivatives and a time derivative. These derivatives appear so often, and field theorists get so tired of writing impossibly long formulas, that they inevitably use a shorthand notation:

$$\frac{\partial \phi}{\partial x} = \partial_x \phi, \quad \frac{\partial \phi}{\partial t} = \partial_t \phi. \tag{4.1}$$

FIELDS

You're supposed to know that these quantities are functions of (x,t), even if the dependence is not written out explicitly. And the spatial derivative $\partial_x \phi$ and the time derivative $\partial_t \phi$, as well as the field value ϕ itself, are what we can use to construct an expression for the energy of the field.

Another consequence of locality is that the total energy in a field configuration (which stretches all throughout space, after all) can be thought of as the integral over space of an **energy density**, the amount of energy per volume of space, defined at each point. Energy density is traditionally denoted by the Greek letter ρ (rho). The total energy in a field configuration is then

$$E = \int \rho(x)\, dx. \tag{4.2}$$

Again, really an integral over all of three-dimensional space, with volume element $d^3x = dx\, dy\, dz$, but for notational convenience we're writing as if we're in just one dimension.

At last we give the expression for the energy density $\rho(x,t)$ of a scalar field:

$$\rho(x,t) = \frac{1}{2}(\partial_t \phi)^2 + \frac{1}{2}(\partial_x \phi)^2 + V(\phi) \tag{4.3}$$

energy density = kinetic + gradient + potential.

There is an obvious resemblance to the energy of a particle. As in the particle case, there is a kinetic energy that depends on "how fast the system is moving," except that now it's how fast the field value is changing at a particular point in space (like a vibrating rubber sheet), rather than how fast a particle is traveling through space. There is also a potential energy that depends on the instantaneous configuration of the system, but again it's the value of the field rather than a particle location. What's new in the field case is the **gradient energy** ½ $(\partial_x \phi)^2$.

This reflects the fact that it takes energy for the field to bend as we move from point to point in space, much like it takes energy for the field to evolve with time.

All of this should make intuitive sense; the behavior of a classical field isn't that much different from the behavior of a stretched rubber sheet. There is kinetic energy from its motion up and down at each point, gradient energy from stretching as it changes from place to place, and potential energy just from the value of the field being displaced from its resting equilibrium value.

There is one feature of the energy-density formula (4.3) that might raise questions. The kinetic energy and the gradient energy both have the number ½ multiplying them. How do we know it's exactly that number? And why is the number the same for both terms?

The answer is relativity. The kinetic energy invokes derivatives with respect to time, and the gradient energy invokes derivatives with respect to space. You might worry that these quantities have different units, but we are implicitly setting the speed of light to unity, $c = 1$, so that time and space can be measured in commensurable terms. The reason why the numbers multiplying the time and space derivatives are the same is that we want our theory to obey the rules of relativity—to be **Lorentz-invariant**, or looking the same in every frame of reference. And for that to happen, in units where $c = 1$, the space and time coefficients have to be equal. The reason why it's ½ is something of a matter of convention, but ultimately it's the same factor that appears in the expression ½ mv^2 for the kinetic energy of a particle.

It's completely possible to invent non-Lorentz-invariant field theories; they might be relevant, for example, if we want to talk about the motion of electrons through a metal. In that case, not all reference frames are created equal: the rest frame of the metal is special. But in this book our interest is going to be in the fundamental behavior of fields for their own sakes, not their particular properties when they

FIELDS

are embedded in some specific environment, so we want to obey the rules of relativity.

FREE FIELDS

There is not much room to play in expression (4.3) for the energy density. The only thing left undetermined is the potential energy $V(\phi)$. Indeed, much of the excitement and richness of particle physics is going to arise from an appropriate choice of potential, not just for a single field but for multiple fields interacting with each other. In the real world, every point in spacetime features a noisy cacophony of fields nudging each other in ways precisely specified by the theory.

But we'll start gently, with a single field and the simplest nontrivial potential we can think of:

$$V(\phi) = \frac{1}{2}m^2\phi^2. \tag{4.4}$$

The parameter m is known as the **mass** of the field. Fields don't really have "mass," of course, as mass is the energy inherent in a localized object at rest, and fields are not localized objects, they're spread throughout space. But upon quantization we'll discover that our field theory describes particles, and m will be the mass of those particles. We just refer to it as the "mass of the field" for convenience.

Why is this the simplest potential? The justification parallels our discussion in *Space, Time, and Motion* of why the simple harmonic oscillator is so ubiquitous. Think about raising the field variable to different powers: $\phi^0, \phi^1, \phi^2, \phi^3$, and so on. ϕ^0 is just a constant; adding it to the energy density doesn't change the dynamics of the field, which cares only about how energy changes from place to place or time to time. ϕ^1 is a linear potential that would push the field in one direction or another. But we want to consider fields that are allowed to just sit there in a minimum-energy state, rather than immediately rolling off.

So we discard that term by hypothesis.* ϕ^2 is a perfectly reasonable thing to consider, so we start with that. We don't need to worry about ϕ^3 and higher-order terms, at least for the moment.

A field whose potential takes on the simple form (4.4) is called a **free field**, because it propagates freely, without interacting with itself or other fields. Indeed, it's the one case in QFT where we can actually solve all the equations exactly. More complicated potentials generate interactions between fields, which we'll delay to the next chapter.

MODES

We're almost ready to quantize our fields. But before we do, let's anticipate a problem. When we considered the quantum mechanics of a single particle with position x, we constructed a wave function $\Psi(x)$, assigning a complex amplitude to each position. Now we are going to have a wave function of field configurations, $\Psi[\{\phi(x)\}]$, assigning a complex amplitude to each possible profile across space. That seems pretty unwieldy. How are we ever going to distill it down to something we can easily interpret, or for that matter calculate with?

The answer is to think about particularly simple field configurations: **plane waves**, vibrations that look like perfectly regular oscillating waves traveling everywhere through space. A plane wave is simple enough that we can describe its behavior exactly.

But plane waves wouldn't be of much interest if they were merely a very special kind of field configuration that was easy to analyze. Their real power comes from an amazing fact: any field configuration at all can be thought of as a combination of different kinds of plane waves.

---

* Somewhat more carefully: let's imagine the potential had a minimum at some value $\phi_{min} \neq 0$. We could just define a new variable, $\phi' = \phi - \phi_{min}$. Then the minimum would be at $\phi' = 0$, so we know there would be no $(\phi')^1$ term in the potential.

FIELDS

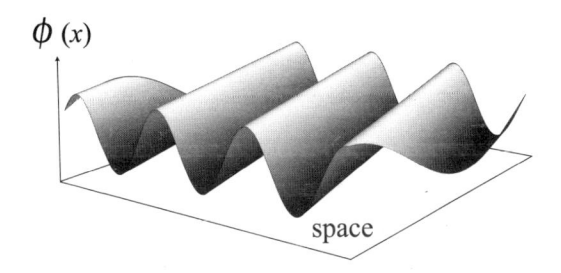

$\phi(x)$

space

That's the magic of Fourier transforms, the mathematical trick we mentioned in Chapter 2 and look at in detail in the appendix. We're going to get a little mathematically specific, but it will be worth the effort.

So rather than thinking of a field configuration as being defined by a separate value at each spatial point $\{\phi(x)\}$, we can think of it as a sum of oscillating waves stretching throughout space, each with a specific wavelength and direction of motion. The individual waves that we add together are called **modes**. The first trick (of many) that is used to make sense of quantum fields is to think mode-by-mode, rather than thinking point-by-point in space. In practice, rather than dealing with the wave function of the whole field, we'll deal with the wave function of a single mode, then put many such modes together. And when we zoom in on the behavior of a single mode, it's just going to be a regular quantum mechanics problem. (In fact, it's going to be a simple harmonic oscillator.)

We're considering a free scalar field, with energy given by (4.3) and potential given by (4.4), and we're going to think about the energy of a single mode with fixed wavelength before combining many modes to make whatever configuration we want. If the wavelength is λ, it's most convenient to characterize our mode in terms of its **wave number** k, which is 2π times the number of wavelengths per distance:

$$k = 2\pi/\lambda. \qquad (4.5)$$

This is convenient because we can write the field configuration of a particular mode as the sum of an oscillating real part and an oscillating imaginary part, as

$$\tilde{\phi}(k,t) = a(t)\cos(kx) - ia(t)\sin(kx) = a(t)e^{-ikx}, \quad (4.6)$$

where $a(t)$ represents the height of this particular vibrating wave. See the appendix on Fourier transforms for more discussion, including the relationship (A.4) between trigonometric functions and the exponential of an imaginary number. Schematically (writing a sum rather than an integral, to keep things conceptually clean), we can combine together modes at each wave number k, with specific heights a_k for each mode, to describe any field configuration at all:

$$\phi(x,t) = \sum_k a_k(t) e^{-ikx}. \quad (4.7)$$

There are some details we won't dwell on right now. There is a separate height function for each mode, which is why we've written the height function in (4.7) as $a_k(t)$. But there are enough subscripts floating around already that we'll suppress this one from here on out, and just write $a(t)$. Classically, we can solve for the actual form of $a(t)$; it oscillates in time with a frequency that is fixed by the properties of the field. Finally, in three dimensions, we would talk about the **wave vector** $\vec{k}$, of which the wave number is its magnitude: $k = |\vec{k}|$. Then a mode would look like $\phi(\vec{k}) = a_{\vec{k}}(t) e^{-i\vec{k}\cdot\vec{x}}$, where $\vec{k}\cdot\vec{x}$ is the dot product between the wave vector and the position. But we'll continue to simplify our lives by pretending that space has only one dimension, at least as far as our notation is concerned, so our modes will each have spatial dependence e^{-ikx}.

The reason why quantum field theory might seem intimidating, even if you have made your peace with the basics of quantum mechanics, is that there are an infinite number of things to keep track of in

FIELDS

the wave function of the field configurations, $\Psi[\{\phi(x)\}]$. There are a lot of field configurations out there. How do we keep track of them all? You might think we can just focus in on a single point in space x_0, study the behavior of the quantum field $\phi(x_0)$ at that point, and then generalize to all spatial points. But that doesn't quite work; because of the spatial derivative terms (the gradient energy) in (4.3), what happens at one point depends on what's happening at all the nearby points in a nontrivial way.

Modes come to our rescue. The power of the Fourier transform is that we can look at one mode at a time, rather than one spatial location at a time. And in that case the spatial dependence of the field is completely taken care of—it's e^{-ikx}, as given in (4.6). So all we have to concentrate on is the wave function of a single number, the height a. That shouldn't be any harder than ordinary quantum mechanics, when we were interested in the wave function of a single position x. In both cases we have a single number (position of the particle, height of the mode) on which our wave function will depend. For particles we have $\Psi(x)$, for a single mode of a free field we will have $\Psi(a)$, but the manipulations will look much the same.

THE ENERGY OF A MODE

Now let's plug the explicit form (4.6) for a single mode of our field into equations (4.3) and (4.4) for its energy density. First note that the derivative of an exponential function brings down whatever is multiplying the variable in the exponent, so

$$\frac{\partial}{\partial x} e^{-ikx} = -ik\, e^{-ikx}. \tag{4.8}$$

The other thing to note is that after the Fourier transform, all of the squared quantities in the energy density (4.3) turn into absolute-values-squared. So all of those sinusoidal exponentials go away, for example:

$$\left|e^{ikx}\right|^2 = e^{ikx} \cdot e^{-ikx} = 1. \tag{4.9}$$

When all is said and done, the energy density of the mode (4.6) looks like

$$\tilde{\rho}(k) = \frac{1}{2}(\partial_t a)^2 + \frac{1}{2}a^2 k^2 + \frac{1}{2}a^2 m^2$$

$$= \frac{1}{2}\left(\frac{da}{dt}\right)^2 + \frac{1}{2}(k^2 + m^2)a^2. \tag{4.10}$$

(We can replace the partial derivatives in the first term with an ordinary derivative, since a only depends on one variable, t.) When we say energy "density," we're now talking about the energy per mode k, not per spatial location x. Physicists will often talk about describing fields "in Fourier space," rather than "in real space." That means we are thinking of the field as a collection of plane wave modes rather than as a collection of values at points in ordinary space. Completely equivalent, but the mode description is more convenient for quantizing the theory.

Okay, now, don't let your eyes glaze over. The little calculation we just did to derive (4.10) for the energy density of a mode with wave number k is one of the most important ideas in quantum field theory, and therefore one of the most important in modern science. Maybe in human history.

To see why, forget everything we've just said about QFT for a moment, and think about a single particle in a simple-harmonic-oscillator potential. Its energy is

$$E_{\text{SHO}} = \frac{1}{2}\left(\frac{dx}{dt}\right)^2 + \frac{1}{2}\omega^2 x^2. \tag{4.11}$$

Stare at this for a bit. Now stare at (4.9). Then stare back here.

What you see is: they are the same equation. The only difference is some re-labeling. Instead of the position of the particle $x(t)$, we have the height of the oscillating mode, $a(t)$. And the role of the angular frequency ω is being played by $\sqrt{k^2 + m^2}$. But the mathematical structure is the same.

This is the crucial insight: if we take a free scalar field and decompose it into a set of modes of wave number k, each individual mode acts exactly like a simple harmonic oscillator with frequency

$$\omega^2 = k^2 + m^2. \tag{4.12}$$

The frequency with which a mode will oscillate depends on both its wave number and the mass of the field. A field whose mass equals zero can oscillate arbitrarily slowly (when k is very small), but mass puts a lower limit on the frequency. And as the wave number gets larger (corresponding to smaller wavelengths), the frequency goes up.

This kind of mathematical prestidigitation is common in physics. We are familiar with a simple problem: a single particle in a harmonic-oscillator potential. In the back of our minds there might be a tangible physical system, maybe an electron trapped in an electric field, or maybe a weight hanging from a spring. But the equations don't care about which system we are thinking of. They represent the essence of the dynamics stripped down to its most basic form, which can be common to many distinct systems. Then we turn to an utterly different physical situation: a plane wave configuration of a free scalar field, oscillating up and down in time. Amazingly, the equations turn out to be the same (with some simple re-labeling). So our work is already done; we know what the solutions look like. In this case, we once again find quantized energy levels.

If (4.12) looks somewhat familiar, think back to the discussion of relativistic energy and momentum in *Space, Time, and Motion*. There we found that the energy and momentum and mass of a particle are related by $E^2 = p^2 + m^2$. This is no coincidence. Once we squeeze some particles out of our quantum fields, the energy of the particle associated with this kind of mode will be $E = \hbar\omega$ and the momentum will be $p = \hbar k$, where $\hbar$ is Planck's constant (which has been invisible since we've been setting it, and the speed of light, equal to 1). This is a general set of relations that are worth keeping in mind. Remembering that physicists label short-distance phenomena as ultraviolet or "UV" and long-distance phenomena as infrared or "IR," we have the following correspondences:

UV	IR
Short wavelength	Long wavelength
Short times	Long times
High energy	Low energy

The moral of the story is that a free scalar field can be thought of as an infinite collection of harmonic oscillators. It's not that there's an oscillator at each point in space; the "oscillators" are individual modes that spread throughout space and can be added together to describe any field configuration we like. I told you that the simple harmonic oscillator was the best spherical cow out there.

WAVE FUNCTIONS OF FIELDS

Now we are ready to quantize our fields. For reasons of conceptual clarity we will set up the problem in terms of field configurations $\{\phi(x)\}$, and then in the next section we'll switch to using modes to make our life easier.

When we talked about entanglement, we emphasized that the wave function is not a field, at least when there is more than one particle in the universe. Fields are things that have a particular value attached to each point in space, whereas amplitudes of the wave function are attached to configurations of whatever physical system we have in mind. If that system is several particles, the configuration space is several copies of physical space, one for each particle, so we have $\Psi(x_1, x_2, \ldots x_N)$ for N particles.

If the wave function is not a field, but we're doing quantum field theory, what is the relationship between the two? The answer is that we have a wave function *of* field configurations, in exactly the same sense that a quantum theory of particles features a wave function of particle positions. That is, in a particle theory we take the set of all possible locations of a particle (or set of several particles), and to each one we assign a complex number. The set of all such assignments is the wave function, defining the quantum state. In a field theory, we take the set of all possible configurations of the field, and to each whole configuration we assign a complex number. The field wave function is the set of all such assignments, which we can write as $\Psi[\{\phi(x)\}]$. Note that $\Psi[\{\phi(x)\}]$ is not a function of x—it's a function of the whole configuration $\{\phi(x)\}$, which is defined at all spatial points. The notation is awkward, but we do the best we can.

This abstract nonsense is a bit easier to swallow in the particle case because we can imagine observing the particle at different positions in space, and the probability of getting that measurement outcome is the corresponding wave function squared. In the field case we have to stretch our imaginations a bit and imagine simultaneously measuring the field at every point in space. Not realistically practical, but it helps us define what we're talking about. The probability of observing a certain field configuration is

$$P[\phi(x)] = \left|\Psi[\{\phi(x)\}]\right|^2. \tag{4.13}$$

Again, this is not the probability of measuring $\phi(x)$ to have a particular value at some particular point x; it's the probability of an entire configuration $\{\phi(x)\}$ over all of space.

There is also a version of the Schrödinger equation that tells us how the wave function evolves with time. All we have to do is to write down the appropriate Hamiltonian operator that characterizes the energy of the state, then we can solve $\widehat{H}\Psi = i\partial\Psi/\partial t$. But just as there are equivalent alternatives to the Newtonian formulation of classical mechanics (Hamiltonian mechanics, Lagrangian mechanics), there are also alternative ways of characterizing the wave functions of quantum states, and those will turn out to be more useful for QFT.

PARTICLES FROM FIELDS

Now that we've talked abstractly about what a "wave function of a field" is supposed to mean, we can get down to brass tacks. The good news is, once we change variables from field configurations $\{\phi(x)\}$ to plane wave modes $\{\phi(k)\}$, actually quantizing our field is a piece of cake. The answer is going to be that the quanta arising from fields can be interpreted as sets of particles.

The free field, which we can think of as a sum of modes, is a collection of simple harmonic oscillators, one for each mode with a given wave number k. Think of the modes one at a time. Essentially we care about the wave function of the single height variable a for each mode k, which we've just established behaves like a harmonic oscillator. The wave function of this mode, $\Psi(a)$, will therefore have the distinctive set of solutions we saw for the oscillator back in Chapter 1. There is a lowest-energy **ground state**, and then a tower of **excited states** with evenly spaced energies above that of the ground state.

The set of quantum states of the entire QFT is just the collection of all of these harmonic-oscillator states for each wave vector k. There will be one unique state of overall lowest energy, the one where all of the individual oscillators are in their ground states. That is called the **vacuum state** of the theory. "Vacuum" is used by physicists not to

FIELDS

mean "empty space" (although that's what it will look like) but rather "the lowest-energy state of a theory."

There is only one way to have zero energy, but there are many ways to have nonzero energy. Starting with the vacuum, we can imagine "bringing to life" just a single mode with wave number k. Since modes are like harmonic oscillators, the quantum state can be thought of as a superposition of that mode's energy levels. And a more general quantum state of the field as a whole will be a superposition of such states for all of the infinite number of modes of various k's. That's a lot to juggle, but we can build up slowly to get a handle on what's going on.

Think of a state in which just one mode is involved, the mode with $k = 0$. Since $k = 2\pi/\lambda$, that's a mode with "infinite wavelength"—basically the field has a constant value everywhere. Classically, that value can oscillate with time; quantum mechanically, it will have a wave function with some profile $\Psi(a)$. Imagine that this mode is in its first excited state. From (4.12) the frequency of the associated harmonic-oscillator potential is just $\omega = m$. The energy of the state is therefore $E = m$, since $E = \omega$, and the momentum is $p = k = 0$. (Remember, $\hbar = 1$.) That's interesting: it's just the energy and momentum of a single particle of mass m sitting at rest.*

Now consider a single mode with some other wave number k, also in its first excited state. It will have energy $E = \omega = \sqrt{k^2 + m^2}$, exactly as we expect for a single particle with momentum $p = k$. Its spatial profile is not constant but looks like a plane wave, e^{-ikx}.

---

* Annoying detail: the formula (1.14) for the energy levels of a simple harmonic oscillator includes a constant contribution ½ $\hbar\omega$. This is the zero-point energy, and we're going to ignore it right now. That's okay, because what physically matters is the differences in energy between different states, not the overall energy. But when we think about gravity, this zero-point quantum field energy becomes the "cosmological constant"; that's an issue we will briefly touch on later.

What if we're interested in field configurations that are not as simple as a single plane wave mode? No problem: we can combine the modes of different k's, all of them in their first excited states, to get any field profile we like. A Fourier transform takes an arbitrary field profile and expresses it as a sum of plane waves, but we can also go backward, combining a collection of plane waves into whatever shape we care about. So for example, we can make a **wave packet**, which is what physicists call a localized wave that oscillates near some particular position in space but goes to 0 far away.

wave packet

+ infinitely more plane waves

What we're seeing is that a quantum state constructed from a superposition of free-field modes in their first excited states looks and acts like the wave function of a single particle of mass m. We didn't put particles in; we started with a field and quantized it. The solutions to the resulting equations are particle-like quanta. This is a case where if it looks like a duck and quacks like a duck, we are justified in declaring it to be a duck. The first excited state of a free quantum field can be interpreted as the quantum state of a single relativistic particle.

When describing the energy of the field in (4.3), we introduced the potential $V(\phi)$, but that's a way of thinking about the energy that is associated with field value (rather than with its derivatives). The particle that has emerged after field quantization isn't trapped in any kind of potential; it's moving freely through space. In the classical limit, it would be moving at a constant velocity in a straight line. It's the height of the mode whose energy involves a harmonic-oscillator potential and therefore has discrete energy levels. We're making clever use of a mathematical equivalence between the potential energy of an actual harmonic oscillator and that of the height of a field mode.

The second excited state of our free quantum field—made from combining the second excited states of various modes—will have a minimum energy $2m$. The third excited state will have a minimum energy $3m$, and so on. That's because they correspond to states that look like two particles, three particles, and so on.

Looking at our QFT in terms of modes has revealed something amazing: a state of a free QFT can be interpreted as a superposition of collections of different numbers of particles. We can think of this in terms of the Hilbert space of the QFT:

$$\text{QFT Hilbert Space} = \begin{pmatrix} \text{unique} \\ \text{zero-particle} \\ \text{state} \end{pmatrix} + \begin{pmatrix} \text{one-particle} \\ \text{states} \end{pmatrix} + \begin{pmatrix} \text{two-particle} \\ \text{states} \end{pmatrix} + \cdots$$

This description is known as **Fock space** in the literature, after Russian physicist Vladimir Fock. It's not a new kind of thing, just a nice way to think about the Hilbert space of a quantum field theory.

You might think that there is more to being a "particle" than simply having energy levels, and in a sense that's right. In particular, we are used to particles having definite locations in space. But that intuition came from the classical view of particles. In quantum mechanics, particles have wave functions, and localization in space only happens when we measure their position. The same thing is true for the particles described by our quantum field. They will have wave functions. Those wave functions might very well be spread out through space, but we are also free to consider relatively compact wave packets centered on some location. The point is that the kinds of particles that pop out of our quantized fields have all the properties and behaviors that quantum particles are supposed to.

This is all pretty amazing. In the original harmonic oscillator (or electron orbits in an atom, for that matter), the "quantum" nature of quantum mechanics wasn't put in by hand. It came out naturally by solving the Schrödinger equation and finding a discrete set of solutions. Precisely the same thing is happening here. We quantized a field, not a set of particles, but we found out that it ends up looking like a collection of quanta, which we interpret as particles. And it's not just semantics—this is what particles really are, quantized excitations of fields. This is the payoff we get for thinking carefully about modes and their energies.

CREATION AND ANNIHILATION

To be fair, the cow we're studying is extremely spherical indeed. A free field theory—one whose energy is the sum of kinetic, gradient, and potential terms as in (4.3), with the simple quadratic potential (4.4)—is a nice starting point because we can solve all the equations exactly, but it makes for an extremely uninteresting universe. The particles in a free field theory cannot radiate, decay, scatter, or otherwise convert into or interact with each other. To describe any of those important processes, we're going to have to complicate our metaphorical cow by adding interactions, not to mention introducing more than one kind of field.

But we can see, even just from our investigation thus far, that field theory gives us the ingredients to describe particles being created and destroyed, such as we see in radioactivity. It wasn't obvious how to do this in a quantum theory that was based on particles from the start, rather than fields. If I have a quantum state of, say, two particles, the wave function can evolve in all sorts of ways, but it will always describe two particles. Whereas the quantum states of just a single field include states with zero particles, one particle, two particles, and so on. So the creation and annihilation of particles is all subsumed into the dynamics of a single quantum field (or multiple such fields).

If we had some deep-seated aversion to quantum field theory, we might try to allow for processes that changed the number of particles within a particle-based approach. To do that we would have to augment Hilbert space, adding together the appropriate spaces for zero particles, one particle, two particles, and so on. Then we would have to modify the Hamiltonian in some way so as to allow transitions between these pieces, allowing (for example) a single particle to decay into several particles. (In the actual world, a neutron left all by itself will decay into a proton plus an electron plus an anti-neutrino.) In other words, we would have to invent Fock space anyway. Upon further investigation, we would find that we'd actually re-created all of quantum field theory, albeit with an especially roundabout series of moves.

What we're running up against is what physicists sometimes call a **folk theorem**—a result that is believed to be true under a wide variety of circumstances but that can't quite be rigorously proven. The folk theorem in question is this: if you want to simultaneously satisfy the requirements of (1) quantum mechanics, (2) special relativity, (3) locality, and (4) changing particle numbers, then any theory you construct will end up looking like quantum field theory, at least at low energies/long distances. (At short distances we might discover string theory or extra dimensions or some other exotic thing.) QFT, in other words, is incredibly robust: it's a nearly unique way of satisfying a set of simple requirements.

FIELDS OF THE WORLD

Quantum field theory is a rich and complex subject. Many smart people devote their lives to understanding it (and honestly there are less-rewarding ways to spend your life). There are many aspects that contribute to this richness. Let's just highlight two.

First, there are a lot of fields. Each of them has its own special idiosyncrasies. We will consider two broad classes: **fermions**, which correspond roughly to "matter" particles, and **bosons**, which create the force fields we're familiar with. Fermions include electrons, quarks, neutrinos, and more; bosons include photons, gluons, gravitons, the Higgs, and more.* Different fields have different spins, are characterized by different symmetries, and correspondingly come with their own notations. It's quite the menagerie.

Second, fields interact in myriad ways, and calculating the physical effect of those interactions can be a bit of a mess. We might say "two electrons scatter off of each other," but what's really going on is that vibrations in some appropriate fields begin to overlap, which induces vibrations in other fields to which they are coupled, and we literally have an infinite amount of activity contributing to the simplest of processes. Now that we've established that quantizing fields leads to particles, henceforth our time will be devoted to explaining how to think about interactions between them. Such interactions are, after all, what make the universe an interesting place.

---

* The graviton hasn't been directly detected, and possibly never will be, because individual gravitons interact extremely weakly with other particles. We only notice the gravity of the Earth because we're feeling the combined gravitational attraction of 10^{50} atoms. But the basic tenets of quantum field theory and general relativity all but guarantee that gravitons exist. It is unique in being a spin-2 particle.

FIVE

INTERACTIONS

That was quite a revelation we had in the last chapter. We start with the simplest possible classical field theory—a single scalar field $\phi(x, t)$, not interacting with any other fields or even with itself, with a simple potential $V(\phi) = \frac{1}{2} m^2 \phi^2$—and plug it into the rules of quantum mechanics. What we find is a set of states with different energies, which look and behave exactly like collections of particles. There's a vacuum state, states with a single particle, states with two particles, and so on. The original classical theory was as continuous, smooth, and non-discrete as we could imagine, but porting it over into a quantum framework turns it into a theory of quanta that we identify as particles. And, just like the field from which they came, those particles are going to be non-interacting: if we set up an initial configuration describing several particles moving toward each other, each particle is going to continue moving in a straight line without being affected by any other particles.

The underlying mechanism for the emergence of lumpiness from smoothness is precisely the same as why we see discrete energy levels for wave functions of electrons in atoms: when we consider the

Schrödinger equation in an appropriate context, we get a set of discrete solutions. The quantumness of quantum mechanics, including quantum field theory, comes from solving the equations, not from the fundamental nature of the ingredients we use to construct the model.

But we aren't going to be satisfied with a theory that only has a single non-interacting field in it. The real world has different kinds of fields, and correspondingly different particles, and they tend to interact with each other. This is where the real power of quantum field theory becomes manifest: not only describing states with multiple particles but also describing processes in which one set of particles evolves into a completely distinct set of particles. Understanding what the fields are, and how they interact with each other, is ultimately going to account for not only particle physics but also the periodic table of the elements, chemical reactions, and all the way up the ladder to the macroscopic world of our experience.

ORIGINS

You will sometimes hear that there are four forces of nature: electromagnetism, gravitation, and the strong and weak nuclear forces. That's a fine way of thinking, and sometimes we will use that terminology. But it's not very rigorous. In the world of quantum field theory, there is not a clean division between "particles" and "forces." There are only fields, interacting with each other in various ways.

Historically, the need for quantum field theory arose because everyone knew about the electromagnetic field and recognized that it would have to be made fully quantum at some point. (They knew about gravity, too, but Einstein's general relativity has more complicated equations than Maxwell's electromagnetism, so we might as well start with the latter.) As early as 1925, Heisenberg, Born, and Jordan studied a quantum theory of just the electromagnetic field—no charged particles—which is essentially a free field theory and not that interesting. In 1927, Paul Dirac figured out how to add electrons to the theory. This led him to a more careful investigation of how to describe spinning particles like electrons in a relativistic context, which ultimately led to his equation and the prediction of **antimatter** (positrons, the antiparticles of electrons). Antiparticles are denoted by putting a bar over the corresponding symbol, so the proton, neutron, electron, and neutrino are p, n, e, and v, while their antiparticles are $\bar{p}, \bar{n}, \bar{e}$, and $\bar{v}$. Any particle that carries a conserved quantity that can be positive or negative, such as electric charge, will have a corresponding type of antiparticle with the opposite value of the quantity. Since electrons have charge -1, positrons must have charge $+1$. Positrons were discovered experimentally by Carl Anderson in 1935, but at first people didn't believe him. They thought he had seen protons, or that he had just seen electrons but was holding his photographs upside-down and hence misinterpreting how the particles bent in a magnetic field.

Interpreting Dirac's insights in terms of quantum field theory took a bit longer. But some of the basic ideas were there, including the need to describe processes in which the number of particles changed, such as when an electron in an atom drops to a lower energy level and emits a photon. In 1934, Enrico Fermi proposed a field theory that described another particle-changing process: **beta decay**, in which a neutron decays into a proton, an electron, and what we would now call an anti-neutrino. By this point quantized fields seemed like a promising way to go.

Fermi first submitted his paper on beta decay to *Nature*, a leading English scientific journal, which rejected it as being too speculative. Partly due to this experience, Fermi changed his focus from theoretical to experimental physics. He was the leader of the group that created the first self-sustaining nuclear chain reaction, which served as a precursor to the atomic bomb.

SCATTERING

The good news is, once we've been through all of that work to quantize fields and discover that they can be thought of as collections of particles, we can subsequently let our hair down and simply talk as if the world is made of particles. We know that it's really made of fields, and sometimes it will be important to recall this fact, but the *language* of particles is a useful and efficient way of talking about the dynamics of quantum fields, as long as we are well aware of its domain of applicability.

There are a number of physically interesting processes that involve interactions between particles. **Scattering** is the most basic idea: a set of particles comes together, they interact with each other, and another set of particles departs. They could be the same particles, such as when two electrons come together, interact through the electromagnetic force, and push each other apart. But they could also be different. We can consider situations where an electron and a positron (its antiparticle) come together and convert into two photons, which then go their separate ways.

A special case of scattering is **decay**, when one particle spontaneously converts into one or more others. Beta decay of neutrons is one example, but there are countless others. Elements like radium or plutonium are radioactive because the particular combinations of particles constituting their nuclei are prone to decaying. We could also consider **absorption**, in which several particles combine together to make one. That process is relevant in nuclear fusion and elsewhere, but generally rarer than decay.

Not every form of particle interaction is a kind of scattering. When an electron is electromagnetically bound to an atom, it is interacting with the nucleus, but it's not scattering, because that interaction persists over time. Likewise when protons and neutrons are stuck together in a nucleus. Scattering is specifically limited to cases where interactions can be ignored at first because particles are far apart, then they become important as particles approach each other, then we can eventually resume ignoring them again.

That makes our lives considerably easier. The paradigmatic situation we consider in quantum field theory is to start with some initial condition of widely separated particles, let them come together and interact, and use the laws of physics to predict what they will scatter into. The "particles" are actually vibrations in quantum fields, but that's okay.

Because it's quantum mechanics we're talking about here, the prediction will not be deterministic. We'll calculate the amplitude for different conceivable scattering outcomes and then square that amplitude according to the Born rule to obtain the corresponding probability. (*We* won't really calculate anything, but we'll learn what goes into such a calculation.) Happily, such experiments suffice to reveal an enormous amount about the underlying laws of physics. Unhappily, the calculations can be extremely difficult to do, and challenging to understand even if we don't explicitly do them. But we are not deterred.

FEYNMAN'S RECIPE

Throughout the 1940s and '50s a great deal of effort was put into developing quantum field theory, although progress was delayed by World War II. Once the war had ended (and theoretical physicists were largely freed from wartime activities), these efforts redoubled. Numerous technical difficulties arose, centered on the fact that a naïve calculation of basic scattering amplitudes seemed to give infinitely big answers—clearly a sign that some basic understanding was lacking. The history of how these problems were resolved, by physicists such as Julian Schwinger, Richard Feynman, Shin'ichirō Tomonaga, and Freeman Dyson, is intricate and fascinating. As is our custom, we are going to skip past all the complications and missteps and get right to the simplest and most direct way of understanding the final answer, in this case the recipe developed by Feynman to calculate particle processes in terms of fun little diagrams.

Despite some early successes, by the 1940s physicists were far from convinced that quantum field theory was on the right track. (Skepticism in certain circles lingered on into the late 1960s.) Feynman, in particular, wondered whether it might not be better to think of the theory as really describing many individual particles rather than an underlying quantum field. He developed an extremely useful way of thinking about interactions in terms of particles, although it soon became clear that the underlying justification for his rules is best understood in terms of quantum field theory.

Here, in a nutshell, is Feynman's procedure:

- Start with a basic set of allowed particle interactions, such as "an electron and a positron annihilate to form a photon."

- Draw a diagram with particles meeting at a point, called a **vertex**, to represent each basic interaction.

- Specify a scattering process of interest, such as "an incoming electron and positron scatter into an outgoing electron and positron." The important thing is to fix what comes in and what goes out.

- Combine the basic interactions in *all possible ways* that might contribute to the specified process. Each way will correspond to a particular **Feynman diagram**.

- To each diagram, associate a complex number using a specified set of rules.

- Add up contributions from each diagram (in general, an infinite number of them). The result is the **scattering amplitude** from the specified initial state to the final state.

INTERACTIONS

- Square the amplitude to obtain the desired scattering probability.

Interaction vertices representing a basic set of allowed particle interactions form the building blocks of Feynman diagrams. Particles are represented by lines, often with features that help us identify what kind of particle we're talking about—although there are more kinds of particles than kinds of lines, so sometimes we just label them or are supposed to know from experience. When multiple lines intersect at a point, that's a vertex. We can figure out what vertices there are by looking at the underlying dynamics of the theory (typically the Lagrangian). Every vertex is assigned a numerical value, depending on the coupling constant of the interaction in question. Then we assemble them in various combinations to make full Feynman diagrams.

ANTIPARTICLES AND TIME REVERSAL

A paradigmatic example of a Feynman diagram vertex is when an electron and a positron come together to convert into a photon. We can think of this as a process playing out in time, running from left to right in the diagram.

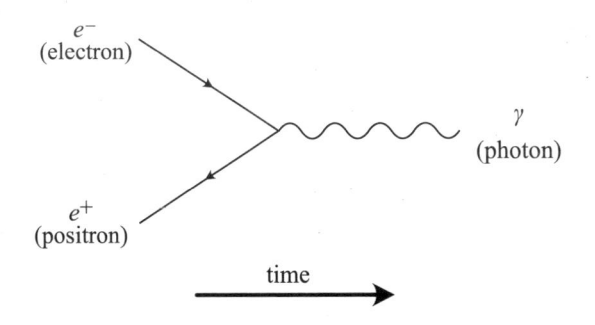

Already we see a number of features that are characteristic of Feynman diagrams. As promised, different particles are represented by different kinds of lines: wavy for photons, and straight for electrons and positrons. (The symbol for a photon is γ, the Greek letter gamma.) The small arrows on the electron and positron lines are crucial: they represent the flow of particleness, as opposed to anti-particleness. So we see that the electron features an arrow that points left to right, in the same direction that time flows in our diagrams, whereas the positron's arrow points the other way because it is an antiparticle. This does not mean that the positron is moving in a different direction; the arrows on the electron/positron lines have absolutely nothing to do with their direction of motion. They simply remind us that the electron is "matter" and the positron is "antimatter." Of course, this is a point of convention: it's not hard to imagine an alternative universe with people made of antiprotons, antineutrons, and positrons, and those antipeople would naturally label positrons as "matter" and electrons as "antimatter." The labels are arbitrary. They are also completely optional. If you want to draw Feynman diagrams where the electron and positron lines don't have any arrows at all, nobody can stop you.

This brings us to another crucial feature of Feynman diagrams: given any interaction vertex, we can construct other vertices (which will be numerically equal once we use these diagrams to calculate amplitudes) by (1) switching the direction in which the particle is moving, and (2) exchanging particleness with anti-particleness. So for example, the previous vertex implies the existence of another vertex in which we move the positron from incoming to outgoing, in the process converting it to an electron. This represents the emission of a photon by a traveling electron.

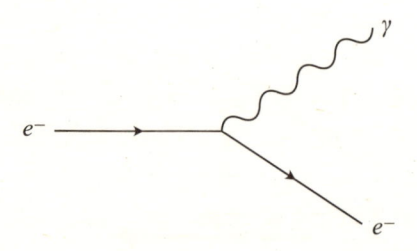

Or we could move both the incoming electron and positron to the outgoing side, as well as the photon to the incoming side. The electron and positron are converted into a positron and electron, but that's the same collection of particles. The photon can be thought of as its own antiparticle, so it remains unaffected. Now the diagram illustrates a photon splitting into an electron/positron pair.

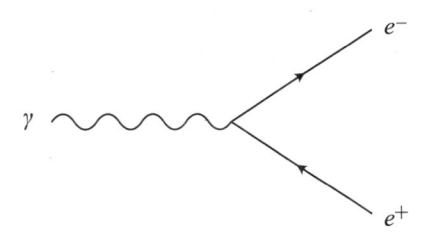

$$\gamma \longrightarrow e^- \,/\, e^+$$

There are also other versions, such as an incoming positron emitting a photon, which you are welcome to play with.

This feature of Feynman diagrams—whenever there is a vertex, there is an equal vertex obtained by switching between incoming/outgoing and matter/antimatter—has led to the motto that "antiparticles are just particles traveling backward in time." That's not utterly wrong, but as with many translations of mathematics into everyday language, it can give you the wrong impression. Nothing "travels" through time, forward or backward; things exist at each moment of time, and the laws of physics guarantee that there is some persistence of things from moment to moment. Certainly any implication that we could use antiparticles to send signals backward in time, or somehow affect the past, is completely off base.

The unimpeachably correct thing to say is that, in Feynman diagrams, the mathematical description of a particle evolving in one direction of time is identical to that of an antiparticle evolving in the opposite direction. All we really care about is the fact that, when you flip a particle line from ingoing to outgoing, the direction of its arrow gets flipped along with it.

AN INFINITE SERIES

Let's put it all together by considering more concretely what we would do to calculate the probability of an electron and a positron scattering off of each other. And to keep things under control we will consider only the electromagnetic force; we won't worry about gravity or Higgs bosons or anything more exotic. Our context is therefore **quantum electrodynamics**, or **QED**, which was the main focus of the 1940s work on scattering in quantum field theory.

You might think, "Well, electrons and positrons are charged particles; they will certainly scatter, so the probability is 1." We're being more particular than that. Our incoming state specifies not only what the incoming particles are but also their individual momenta, and likewise for the outgoing particles. In general there will be a probability that the outgoing particles move in various directions, and we can calculate that.

So we want to construct all possible diagrams that contribute to this process, calculate the numerical contribution from each one, and add them all up.

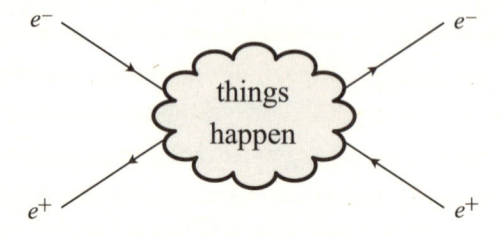

There is one fairly obvious contribution to this process: the electron and positron convert into a photon by the interaction we've already looked at, then that photon converts back into an electron/positron pair. But there is another one we can think of that is just as simple: the electron doesn't annihilate with the positron but simply tosses a photon to it.

INTERACTIONS

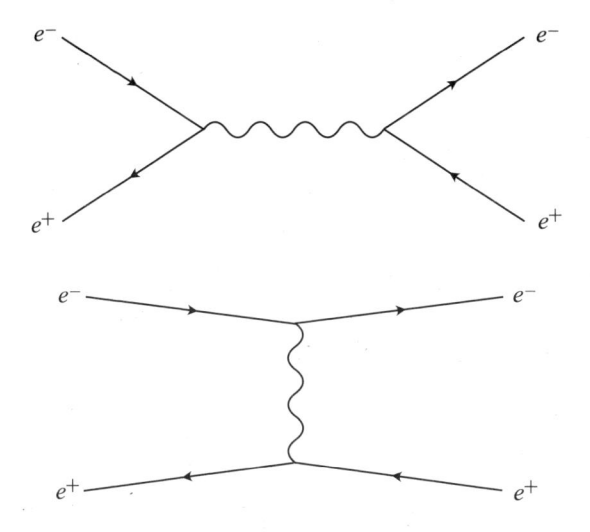

You might worry that it's not clear whether the photon was emitted by the electron and absorbed by the positron, or vice versa. Not to fret; the recipe says we should sum up everything that can happen, so this diagram is meant to encompass both possibilities. Because the photon is its own antiparticle, we don't need to draw separate diagrams for different orientations.

In fact, don't take the diagrams themselves too literally, as if they were representations of pointlike classical particles moving along those trajectories. When we draw Feynman diagrams, the particle lines do not actually represent particular positions in space, but rather particle states with definite momentum. And such states are completely delocalized in space, according to the uncertainty principle. That's fine, since the diagrams are just devices to help us do calculations in quantum field theory, not photographs of physical processes. You can think of the result as representing the probability of this process occurring if the particles are right on top of each other; we can always correct things later to account for particles passing by at some degree of separation.

And of course we can add in more complicated diagrams, as long as we stick with our basic vertex. So, for example, we'll never have two electrons converting into a photon, or an electron/positron pair converting into a single electron—there are no such vertices. There are infinitely many possibilities, and our task is to add them all up.

Feel free to draw all the diagrams you can think of. There is an endless supply of them, you won't run out.

CALCULATING AMPLITUDES

The point of these diagrams isn't merely to provide suggestive visualizations of what can happen when particles interact with each other. They are the basis for precise numerical predictions of scattering probabilities. This is where the complete mathematical formalism gets complicated, and a typical graduate student in physics spends a year or so of their education coming to terms with it. In our quest to focus on the important underlying ideas, we're going to gloss over a textbook's worth of equations to present the essence of the process.

That essence is: each interaction vertex is associated with a number, the **coupling constant** corresponding to that interaction. The contribution from any particular diagram is calculated in two steps:

1. Multiply the coupling constants associated with every interaction vertex appearing in the diagram.
2. Add up contributions from all possible energies and momenta of particles appearing in the interior of the diagram.

It's the second step here that leads to infinities and other conceptual problems; we will give it proper consideration in the next chapter. For now, let's think about the first step.

In QED, the coupling constant associated with the fundamental vertex is $\sqrt{\alpha}$ where α (Greek letter alpha) is the fine-structure constant, a dimensionless number (that is, one with no units like length or time) that experiment measures to be

$$\alpha \approx \frac{1}{137}. \tag{5.1}$$

The reason why we choose a convention such that the square root of α, rather than α itself, is associated with the vertex is that real-world processes, such as electron/positron scattering, generally have two or more vertices in them. (An electron and a positron cannot annihilate into a single, real photon while conserving energy and momentum.) So the simplest diagrams that contribute to this scattering process are proportional to α.

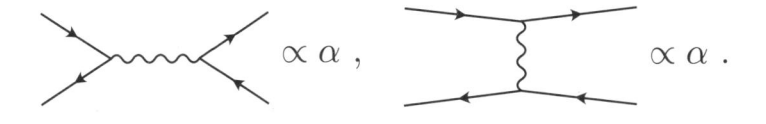

$\propto \alpha$, $\quad \propto \alpha$.

More complicated diagrams, with more and more vertices, will naturally contribute higher and higher powers of α. We say "proportional to" rather than "equal to" because we're neglecting numerical factors like 2π, as well as factors that depend on the momenta of the

$$\propto \alpha^2, \qquad \propto \alpha^2.$$

incoming particles. Those all matter crucially for the working particle physicist, but we want to isolate the basic physical principles.

This dependence on α explains, at last, why there is at least a hope that the sum of an infinite series of diagrams could give us a finite answer: because more complicated diagrams contribute much smaller amounts to the ultimate answer. That's because α is a small number, much less than 1. It's something like considering the infinite sum

$$\frac{1}{2}+\frac{1}{4}+\frac{1}{8}+\frac{1}{16}+\frac{1}{32}+\cdots=1. \qquad (5.2)$$

Just because we are adding up infinitely many numbers, it doesn't mean the answer can't be finite and well defined, as long as those numbers get sufficiently small as we travel down the sequence. It's not obvious whether it will or won't be—even though contributions from more complicated diagrams are quite small, there are a lot more diagrams to consider, so it might not be clear whether the series "converges" to a finite answer. In fact, after a great deal of work, it's possible to show that the Feynman diagram series in QED almost converges, but not quite. What this means in practice is that you can't get an exact answer this way, but the first few terms (the lowest powers of α) actually give quite a good approximation to real-world processes.

The Feynman diagram series for QED is an example of **perturbation theory**, which we mentioned in *Space, Time, and Motion*. That's a technique by which we solve a hard problem (interacting fields in quantum electrodynamics) by first solving an easy one (free fields, with interactions turned off) and then gradually correcting that answer in terms of powers of a small perturbation. In QED it's the fact that the fine-structure constant is so small that makes perturbation theory so powerful. Sometimes, as in the interactions of quarks and gluons inside a proton, the relevant coupling constants aren't small at all, and we have to use different techniques. But it's a fortunate feature of our physical universe that many interactions can be thought of as providing small perturbations to the non-interacting dynamics of the system we are considering.

LAGRANGIANS FOR FIELDS

Feynman diagrams provide an elegant way to calculate the probability of particle-scattering events within the framework of quantum field theory. But so far we've been talking about the diagrams in their own right, without making much contact with the underlying field theory. That contact comes from the precise way in which numerical values, such as the fine-structure constant, are attached to diagrams. And that is best understood by thinking about the Lagrangian approach to field theory. The connection will turn out to be quite simple and elegant: each term in the Lagrangian is directly associated with part of a diagram, either the lines representing moving particles or the interaction vertices where they meet.

In *Space, Time, and Motion* we defined the **Lagrangian** in classical mechanics as the kinetic energy minus the potential energy, $L = K - V$, and the **action** of a trajectory as the integral of the Lagrangian over time, $S = \int L \, dt$. Considering every possible trajectory (or "history") of a system between some starting point and some ending point, the one that actually happens—the trajectory that obeys the laws of classical physics—is the one that has the minimum action of them all. For a particle, that trajectory is position as a function of time, $x(t)$, while for a field it's an entire spacetime profile, $\phi(x, t)$, but the essence of the idea is the same in both cases.

The quantum version is a little different, based on Feynman's **path integral** or "sum over histories." Now all of the paths count, rather than just the one with least action S. Rather than looking for one special path with minimum S, we calculate the complex number e^{iS} along every path, then integrate that over all paths to obtain the quantum amplitude for starting in some configuration and ending in another.

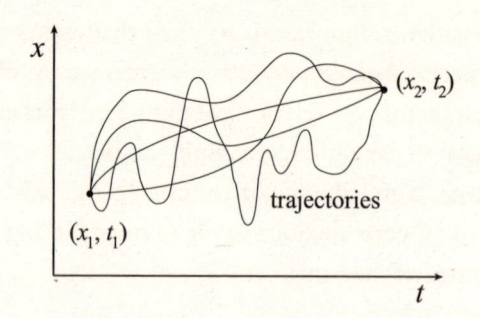

That works because of a property we might remember about calculus: at the minimum of a function, the derivative is 0. That means that values of the function near the minimum are almost equal to each other. So in the vicinity of the least-action path, the values of e^{iS} are approximately the same; when integrating that quantity over all the paths, they add up and reinforce each other, giving us a large probability. Away from the path of least action, nearby paths give very different values for e^{iS}, positive and negative as well as positive-imaginary and negative-imaginary, so they interfere and cancel out. That's how, from the path-integral perspective, classical-looking behavior is likely but not inevitable.

When we talked about the energy of a field in the previous chapter, we noted that locality implies a particular way of calculating that energy: there is an energy density $\rho(x)$, constructed from the field and its derivatives, and we integrate it over space to get the total energy, $E = \int \rho(x)\,dx$. The same thing is going to happen for the Lagrangian: we will construct a **Lagrange density** $\mathcal{L}(x)$, then integrate over space to get the actual Lagrangian,

$$L = \int \mathcal{L}(x)\,dx. \tag{5.3}$$

INTERACTIONS

As usual we are writing a single coordinate x, but you should think of that as standing in for all three spatial coordinates. The action is still the integral over time of the Lagrangian, so it's the integral over all spacetime of the Lagrange density: $S = \int L \, dt = \int \mathcal{L} \, dx \, dt$.

When particle physicists talk about a specific theory or model, they usually are thinking of defining it by writing down the Lagrange density. Many of the working hours in the lives of theoretical particle physicists for the last half century have been spent thinking about Lagrange densities. So much so that we are usually lazy and just say "the Lagrangian" when we really mean "the Lagrange density," assuming you are sophisticated enough to know that you have to integrate it over space to get the actual Lagrangian (and over time to then get the action).

A Lagrangian is supposed to be kinetic energy minus potential energy. For fields, we have a new contribution: the gradient energy. Fortunately, the requirement that we be compatible with relativity (space and time are two different aspects of the underlying spacetime) forces there to be a definite relationship between kinetic (time derivative) energy and gradient (spatial derivative) energy. As a result, the gradient energy comes in with a minus sign compared to the kinetic energy. (It's the same minus sign that appears in the metric tensor on Minkowski space.) So at this point we have:

$$\begin{pmatrix} \text{Field} \\ \text{Lagrangian} \end{pmatrix} = \begin{pmatrix} \text{Kinetic} \\ \text{Energy} \end{pmatrix} - \begin{pmatrix} \text{Gradient} \\ \text{Energy} \end{pmatrix} - \begin{pmatrix} \text{Potential} \\ \text{Energy} \end{pmatrix}$$

In the free-scalar field theory we discussed in the previous chapter, for example, the Lagrangian is

$$\mathcal{L} = \frac{1}{2}(\partial_t \phi)^2 - \frac{1}{2}(\partial_x \phi)^2 - \frac{1}{2}m_\phi^2 \phi^2. \tag{5.4}$$

For this free field theory, we could solve all the equations exactly. Modes of definite wavelength act like simple harmonic oscillators, and we can interpret excitations of the field as freely moving particles. If we have several free fields, we can just add their Lagrangians together. The result describes several kinds of particles, but they still travel without interacting.

Now we want to introduce interactions. That's where all the interesting stuff is going to happen. We assume we can solve the noninteracting theory exactly, and then we include interactions and treat them using perturbation theory, generating an infinite sequence of Feynman diagrams. A more useful decomposition is therefore:

$$\begin{pmatrix}\text{Field}\\ \text{Lagrangian}\end{pmatrix} = \begin{pmatrix}\text{Free}\\ \text{Lagrangian}\end{pmatrix} + \begin{pmatrix}\text{Interaction}\\ \text{Lagrangian}\end{pmatrix}$$

Almost always—not absolutely always, but enough that we won't worry about the exceptions here—interactions are going to be defined by additions to the potential energy, not the kinetic/gradient energy. Notice that all of the free-field energy densities—the kinetic energy ½$(\partial_t \phi)^2$, the gradient energy ½$(\partial_x \phi)^2$, and the kinetic energy ½$m^2 \phi^2$—feature two appearances of the field variable ϕ. The kinetic energy has the time derivative of the field squared, the gradient energy has the spatial derivative of the field squared, and the potential energy has the field itself squared. Interactions are going to be given by multiplying three or more field variables together. Here, for example, is a possible Lagrangian describing two interacting scalar fields, ϕ and θ:

$$\begin{aligned}\mathcal{L} = &\frac{1}{2}(\partial_t \phi)^2 - \frac{1}{2}(\partial_x \phi)^2 - \frac{1}{2}m_\phi^2 \phi^2 \\ &+ \frac{1}{2}(\partial_t \theta)^2 - \frac{1}{2}(\partial_x \theta)^2 - \frac{1}{2}m_\theta^2 \theta^2 \\ &- A\phi^2 \theta - B\phi^2 \theta^2.\end{aligned} \quad (5.5)$$

The first three terms on the right-hand side (the first line) represent the free Lagrangian for ϕ, with m_ϕ representing its mass. The

INTERACTIONS

next three terms (second line) are the free Lagrangian for θ, with m_θ representing its mass, which might very well be different from the mass of ϕ. All of those terms include two appearances of the appropriate field. The final two terms (last line) are the interactions, with A and B playing the role of coupling constants. The A term has three fields in total (ϕ twice, θ once) and the B term has four fields in total (ϕ twice and θ twice).

Now we just have to see how those interactions turn into vertices in diagrams.

FROM LAGRANGIANS TO DIAGRAMS

Think about a single free field ϕ. What would its Feynman diagrams look like? There aren't any interactions, and therefore no vertices. The particle just propagates from place to place, undisturbed. But we could, if we really wanted to, represent this diagrammatically. It would simply be a single straight line, representing a propagating ϕ particle.

$$\phi \longrule \phi$$

Not very exciting, really. For more complicated fields, including electrons and photons, this **propagator** will actually have a nontrivial mathematical form. One way or another, it just represents the idea that the particle travels unimpeded from one place to another.

But notice something about the propagator diagram. For a free field, all of the terms in its Lagrangian have two appearances of the field. And the propagator has two ends, one going in and one coming out, both labeled by that field.

Let's make a bold generalization: every term in the interaction Lagrangian is associated with a vertex with one line for each field that appears in the term, and the numerical value associated with it is simply its coupling constant. For the two-field model (5.4), for example, there are two interaction terms, $A\phi^2\theta$ and $B\phi^2\theta^2$. Let's denote ϕ by a solid line and θ by a dashed line. Then we have two vertices:

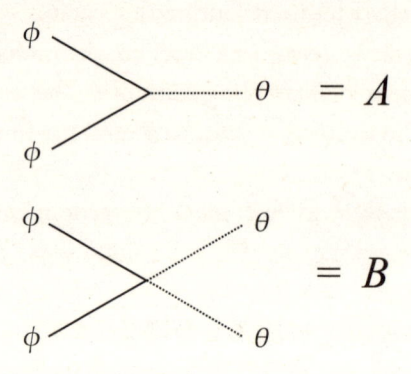

$$\phi\phi\theta \text{ vertex} = A$$

$$\phi\phi\theta\theta \text{ vertex} = B$$

That's basically what you need to know to construct all the Feynman diagrams for this little theory. Feel free to play around drawing some complicated ones. Note that you can only do what the vertices allow you to. A single ϕ never spontaneously changes into a single θ, for example; neither do two θ's annihilate into a single ϕ. But many other interactions are possible. (There is no distinction in this simple theory between particles and antiparticles, since the particles don't carry any conserved quantities that could be either positive or negative.)

A realistic theory like QED will be more complicated, even with just a small number of fields, because the fields themselves will be more complicated geometric objects than just numbers. But the basic idea is the same. If we label the electron field as e^-, the positron field as e^+, and the electromagnetic (photon) field as γ, the QED interaction Lagrangian is essentially

$$\mathcal{L}_{QED} = -\sqrt{\alpha}e^-e^+\gamma. \quad (5.6)$$

This term translates directly into the fundamental QED vertex, where an electron and a positron convert into a photon (or variations thereof).

You can see why particle physicists spend all of their time thinking

INTERACTIONS

about Lagrangians. The Lagrangian defines the dynamics of individual fields as well as how different fields interact with each other. Once you know that, the rest is just turning a crank to make some predictions. Lagrangians imply propagators and vertices, which are put together to make diagrams, which are added up to calculate amplitudes, which are squared to get probabilities.

VIRTUAL PARTICLES

Okay, it's not quite as simple as the previous paragraph might imply. There are subtleties in Feynman diagrams and how they are used. Some of those subtleties are simple and fun; some are challenging and profound. Let's look at the former, saving the latter for the next chapter.

Every Feynman diagram tells a story. Some particles come in, they interact by exchanging various other particles, and another set of particles goes out. But not all of these particles are created equal. All ingoing and outgoing lines represent "real," actually existing particles. But a line that is completely internal to a diagram—beginning on some vertex and ending on some other vertex, never reaching the outside world—represents a **virtual particle**. Virtual particles are not real particles. They represent a real process—the interacting vibrations of a set of quantum fields—but not actual, real-life particles that we could see in an experiment.

The distinction between real and virtual particles becomes relevant because of momentum conservation. In *Space, Time, and Motion* we saw that in relativity, energy and momentum are unified into a single "momentum four-vector," with energy being the zeroth (time-like) component of the vector. For real particles, however, the components of the momentum vector are not all independent; the energy E is related to the spatial momentum $\vec{p}$ and the mass m via

$$E^2 = \vec{p}^2 + m^2. \tag{5.7}$$

When we write down Feynman diagrams, we are thinking of particle states with well-defined momentum. And there is a rule that says *all components of momentum are exactly conserved at every vertex in a diagram*. Scattering dynamics do not violate conservation of energy and momentum.

That has some immediate consequences. For example, the fundamental QED vertex we considered, in which an electron and a positron convert into a single photon, is *not* an allowable Feynman diagram all by itself. That's because there's no way to add up the momentum of two massive, slower-than-light particles (an electron and a positron) and get a result equal to a single massless, speed-of-light particle (the photon). To see this, notice that there is always a frame of reference where the electron's spatial momentum is equal and opposite to that of the positron's momentum, so the total spatial momentum is 0. But a photon can't have zero spatial momentum; that would imply it was sitting still, not moving at the speed of light.

This explains why, when electrons and positrons come together and annihilate, they have to produce two photons, not just one. While our fundamental QED vertex isn't a respectable Feynman diagram all by itself, there's no issue with having it be part of a larger diagram.

Let's look at a process in which an electron and a positron annihilate into two photons. Unlike our previous diagrams, here we have explicitly illustrated the flow of energy and momentum. Doing so is completely optional, depending on your purposes. The incoming electron is 1, the positron is 2, the outgoing photons are 3 and 4, and the virtual particle in between is 5. The momentum arrows have nothing to do with the arrows on the lines that denote the flow of particleness; those are completely different concepts. You might worry that it's ambiguous whether the virtual particle is an electron or positron. The good news is, for virtual particles it doesn't matter, as we ultimately want to include both possibilities.

INTERACTIONS

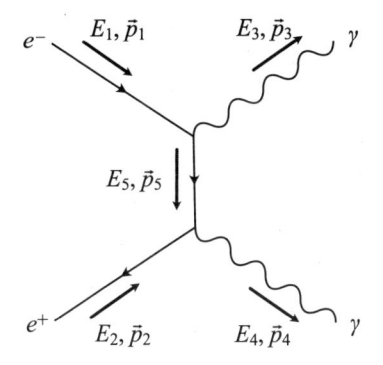

[Feynman diagram: e^- with $E_1, \vec{p}_1$ and e^+ with $E_2, \vec{p}_2$ incoming; virtual particle $E_5, \vec{p}_5$ in the middle; outgoing photons γ with $E_3, \vec{p}_3$ and $E_4, \vec{p}_4$.]

The fact that both energy and momentum are conserved at each vertex gives us the following relations:

$$\vec{p}_1 = \vec{p}_3 + \vec{p}_5$$
$$E_1 = E_3 + E_5$$
$$\vec{p}_2 + \vec{p}_5 = \vec{p}_4$$
$$E_2 + E_5 = E_4. \qquad (5.8)$$

The momentum arrows in the diagram tell you how to add things together: momentum 1 splits into momenta 3 and 5, while momenta 2 and 5 combine to give momentum 4.

By adding and subtracting these equations in appropriate combinations, you can convince yourself that the energy E_5 and momentum $\vec{p}_5$ of the virtual particle are completely determined by the incoming and outgoing energies and momentum; there is no wiggle room. But in general, they will *not* satisfy the classical relationship between energy, momentum, and mass:

$$E_5^2 \neq \vec{p}_5^{\,2} + m^2. \qquad (5.9)$$

Indeed, the energy is just going to be whatever it has to be to satisfy the relations in (5.8); it can even be negative. That is important in explaining how, for example, an electron and a positron can attract each other by "tossing" a photon from one to the other. Shouldn't tossing a photon push you away? Not if the photon has negative energy. Then you actually get pulled in the same direction you toss the photon toward.

And it's all okay, because the particle is virtual, not real. You cannot create a bucket of negative energy by capturing virtual particles; they exist only as mathematical tools inside Feynman diagrams. But they do provide an awfully convenient way of thinking about processes in quantum field theory, and there is sometimes a temptation to forget that they are not what's really happening.

And sometimes, thinking hard about virtual particles leads us to some wild places. That's the subject of the next chapter.

SIX

EFFECTIVE FIELD THEORY

Physics has always had an uneasy relationship with infinity. On the one hand, it appears all over the place, often in extremely useful ways. When Newton and Leibniz developed calculus, part of the challenge was dealing with infinity systematically. And infinity doesn't only refer to things that are "infinitely big"; there are an infinite number of real numbers between 0 and 1. Indeed, any mathematical structure that is smooth and continuous, from spacetime to Hilbert space, has an infinite number of elements—and all of our best current ideas about fundamental physics make use of such structures. And while we don't really know, it's perfectly possible that our real universe stretches infinitely far in space, or time, or both.

On the other hand, we think of ourselves as finite beings, restricted to a finite part of the universe (in time as well as in space), with finite capacities to observe and experience. So if we write down a theory and discover that it predicts that some tangible physical quantity will grow infinitely big right in front of us, we tend to worry. According to general relativity, for example, the interiors of black holes will contain singularities where the curvature of spacetime appears to go to infinity. Most physicists take this as evidence that general relativity, which is a classical theory, after all, just isn't up to the task of describing what happens in these extreme circumstances. In classical statistical mechanics, a straightforward calculation of blackbody radiation seems to predict an infinite amount of short-wavelength radiation—the ultraviolet catastrophe. Thinking about fixing that problem set us on the road to quantum mechanics.

One of the immediate challenges faced by the pioneers of quantum field theory was the recognition that certain terms in the infinite Feynman diagram series, which were supposed to get smaller as the diagrams got more complicated, actually gave infinite answers. Addressing this problem led to the study of **renormalization**, and dramatically changed how we think about quantum field theory. This effort culminated in the idea of **effective field theories**, and this chapter will explain what that means. The basic idea, in brief:

- In standard quantum field theory, when we sum up contributions from virtual particles with arbitrarily high energies, certain diagrams give unphysical infinite results, which need to be somehow subtracted off.

- In the effective field theory paradigm, we deal with this by simply not including contributions from high-energy particles. We introduce an "ultraviolet cutoff energy" Λ, and include only contributions from virtual particles below that energy. That gives us a viable "infrared theory"—one that only refers to particles with energies below the cutoff.

- The resulting effective field theory is defined by a set of effective coupling constants. These "constants" aren't really constant: they depend on the value of the cutoff Λ. And they do so in such a way that none of the eventual physical predictions depend on the value we chose.

EFFECTIVE FIELD THEORY

- Everything in the resulting effective field theory is well behaved and finite, as long as we restrict our attention to particles and processes below the cutoff energy.

The concept of an effective field theory with an ultraviolet cutoff is absolutely central to modern physics. It's one of the biggest ideas we'll explore in this book.

LOOP DIAGRAMS

Let's investigate where those infinities came from in the first place. At the end of the previous chapter we examined a simple Feynman diagram for an electron and a positron annihilating into two photons. Forget all the details of the particular particles and just focus on the idea of two particles coming in and two particles going out, with one virtual particle in between. We could trace the flow of momentum through the diagram and notice that the energy and momentum of the virtual particle were completely fixed by the energies and momenta of the incoming and outgoing particles.

What if we look at a similar situation, but now with more than just a single particle being exchanged? For convenience we'll use p as an all-purpose label for the momentum four-vector, which includes the energy as its zeroth component. When speaking casually, physicists often use "momentum" and "energy" as essentially interchangeable terms in this context—it's usually the momentum four-vector that matters. And the fundamental principle is that momentum is conserved: at each vertex, the sum of the incoming momenta must equal the sum of the outgoing.

By staring at the diagram you can figure out what is going on. Reading left to right, we have p_1 and p_2 as momenta coming into the diagram, and p_3 and p_4 as outgoing. At each vertex, the sum of the ingoing momenta equals the sum of the outgoing. The direction in which we draw those momentum arrows is arbitrary; we could reverse them as long as we also multiplied the corresponding momenta by −1 when we add them up.

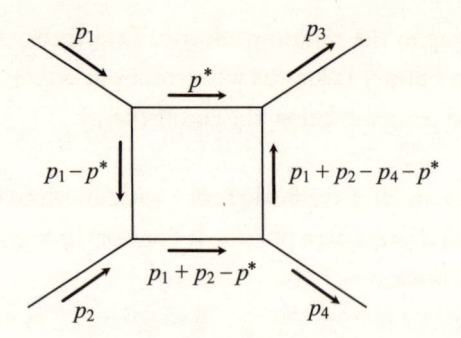

At the first vertex in the upper left, the momentum p_1 gets split into some value p^* going through the upper virtual particle, and $p_1 - p^*$ going vertically downward. Following around the square, at the upper right we have

$$(p^*) + (p_1 + p_2 - p_4 - p^*) = p_3. \tag{6.1}$$

The p^*s cancel each other, and we can move p_4 from the left to the right side to obtain

$$p_1 + p_2 = p_3 + p_4. \tag{6.2}$$

This equation is automatically satisfied, no matter what the individual momenta actually are. It's just conservation of momentum for the scattering process as a whole, equating the sum of the incoming momenta to the sum of the outgoing.

This means that the value of p^*, the initially unspecified momentum going through the top part of the square, could be anything. It is entirely unconstrained, and overall energy-momentum conservation will still be respected, both for the real scattering process as a whole

EFFECTIVE FIELD THEORY

and at every vertex. We think of p^* as the "internal momentum" of the diagram. You see that it appears in the momentum of each internal line.

What do we do about that? The Feynman procedure tells us to add up a contribution from every diagram we can draw. Since any value of p^* would work in our diagram, we should add up all of the contributions for every value of p^*. Really that's doing an integral over all possible values of the internal momentum, but an integral is just a sum over a smooth variable, so we won't distinguish between them.

$$\text{[diagram]} = \sum_{p^*} \text{[diagram with } p^*\text{]} = \infty$$

The issue we immediately run into is that this integral is, for many theories, equal to infinity. That seems bad. More complicated diagrams were supposed to give smaller contributions, since they include more vertices and therefore more coupling constants, but this one (and many even more complicated ones) ends up being infinitely big.

Don't worry, this is a solvable problem. But let's be clear about where the problem arises. It's a matter of the topology of the diagram. In some diagrams, if you start at any point and trace through it, you will inevitably end up at another external line. These are called **tree diagrams**. But in other diagrams, you can find one or more closed loops and circle around forever. These are **loop diagrams**.

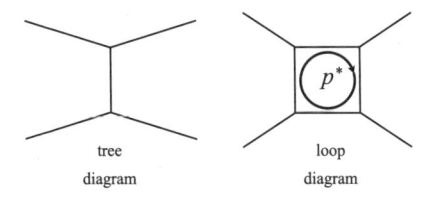

tree diagram   loop diagram

It's the loop diagrams that lead to infinities. For every closed loop there will be an internal momentum (or "loop momentum") that we have to integrate over all possible values, and these integrals generally give infinity. It's a sign that either the theory is simply wrong or we need to be more careful in how we think about what's happening. (It's the latter.)

EFFECTIVE FIELD THEORY

The puzzle of infinities was dealt with by Tomonaga, Schwinger, Feynman, Dyson, and their colleagues, with the first three sharing the Nobel Prize in Physics in 1965 for their efforts. The procedure they invented was dubbed renormalization. It is essentially a method for subtracting off infinity in a particular way that allows us to express the final answer to a scattering calculation in terms of physically measurable, finite quantities.

Not everyone found it to be a completely satisfying procedure. Paul Dirac remained skeptical throughout his life: "Sensible mathematics involves disregarding a quantity when it is small—not neglecting it just because it is infinitely great and you do not want it!" Feynman himself called it "a dippy process," "hocus-pocus," and suspected that it was "not mathematically legitimate."[*]

Happily, we're not going to spend time fretting over the mathematical legitimacy of old-fashioned renormalization theory, because we can appeal to a more modern and well-formulated approach: effective field theory (EFT). As pioneered by physicist Ken Wilson and others in the 1960s and '70s, the EFT paradigm provides a systematic way of calculating observable quantities without first conjuring up infinities and then subtracting them off.

---

[*] Dirac is quoted in H. Kragh, *Dirac: A Scientific Biography* (Cambridge University Press, 1990), 184. Feynman's quotes are from R. P. Feynman, *QED: The Strange Theory of Light and Matter* (Penguin, 1990), 128.

EFFECTIVE FIELD THEORY

Wilson was inspired by a newfangled device that was beginning to become useful to physicists: the digital computer. Imagine we try to do a numerical simulation of a quantum field theory. Inside a finite-sized computer memory, we cannot literally calculate what happens at every location in space, since there is an infinite number of such locations. Instead, we can follow the fields approximately by chunking space up into a lattice of points separated by some distance. What Wilson realized is that he could systematically study what happens as we take that separation distance to be smaller and smaller. And it's not just that we're doing calculus, taking the limit as the distance goes to zero. Rather, at any fixed lattice size, we can construct an "effective" version of the theory that will be as accurate as we like. In particular, we can get sensible physical answers without ever worrying about—or even knowing—what happens when the distance is exactly zero.

The insight that starts us down the road to effective field theory is: be honest about what you know. Think about precisely why we were getting infinite answers in the first place. It was because we were adding contributions from loop momenta all the way up to infinity. But in field theory there is an inverse relationship between energy/momentum and space/time, so that any particular momentum p is associated with a distance λ (in units where $\hbar = c = 1$) according to

$$p \sim \frac{1}{\lambda}. \tag{6.3}$$

So very large momenta correspond to very small distances. When we include loop momenta that range up to infinity, we're including contributions from virtual particles with arbitrarily high energies, at arbitrarily small distances. What right do we have to do that? We can't claim to know how physics works in those conditions. There could be all sorts of new particles and forces that we don't know about. Maybe spacetime itself ceases to be a useful concept once distances become incredibly small. We should be more up-front about our ignorance and formulate answers that don't depend on any such assumptions.

Therefore, Wilson proposed, be brutally honest. Let's not add loop momenta up to infinitely big energies, but only up to some chosen **ultraviolet cutoff**, Λ. (Remember that "ultraviolet" or "UV" is shorthand for large energy/short distance, just as "infrared" or "IR" is small energy/long distance.) That is, we introduce an energy cutoff Λ as a new parameter in the theory and divide the space of particle momenta into UV (above Λ) and IR (below Λ). Then we simply don't include any of the UV contributions when we do our loop integrals. Remember that energy is related to momentum and mass by $E^2 = p^2 + m^2$, so UV particles are those with either high mass or high momentum (or both). Virtual particles with short wavelengths are left out of our consideration entirely, and we only talk about IR particles.

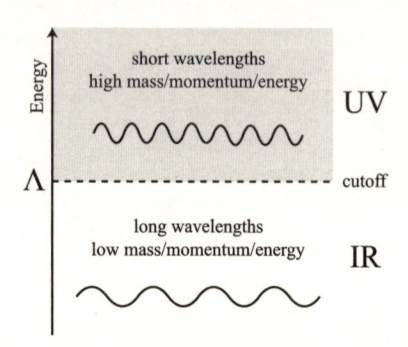

It's not too hard to see that this brute-force procedure will get rid of the infinities. We're summing up a finite number of contributions (for each diagram), each of which is itself finite, so we'll get a finite answer. But we might worry that the resulting formalism just yields nonsense, since we've left out so much. Amazingly, the dynamics of the IR particles, all by themselves and with any trace of UV particles removed by hand, can be a perfectly good quantum field theory, just one with different coupling constants than the theory we started with. That's why such theories are called "effective." This is remarkable because in physics (or science more generally) you don't typically get to excise certain parts of your theoretical description and expect to be left with anything sensible at all. It doesn't make sense to think in terms of "classical mechanics but we ignore momentum," for example.

This is a good news/bad news situation. The good news is, maybe we don't know what happens in the UV, but we don't need to know; what happens there might influence the IR, but all such effects can be bundled up in a small number of parameters that can be specified in IR terms alone. Particles will count as UV if they have mass $m > \Lambda$, or if they are low mass but are moving rapidly and have large momentum. So the potential existence of either unknown heavy particles or new short-wavelength phenomena (new forces, maybe even a breakdown of spacetime itself) doesn't stop us from describing what happens in the IR world of low-mass, low-momentum particles.

The bad news is that this makes it hard to experimentally probe what's happening in the UV, precisely because it doesn't matter very much to what we see in the IR. That's why particle physicists need to build gigantic accelerators to find new particles. To figure out what's happening in the UV, we have to literally visit the high-energy UV regime. Our current best accelerators smash particles together with energies of order 10 TeV (trillion electron volts). It's extremely hard for us to get direct experimental data that would tell us what's going on at even higher scales, but perhaps future colliders will get there.

The other remarkable property of Wilson's idea is that, despite the fact that we had to introduce a seemingly arbitrary ultraviolet cutoff Λ, the value of that cutoff is going to completely disappear from our physical predictions. The IR theory is self-contained and independent of the actual cutoff we use. The secret is that the coupling constants defining our theory will depend on Λ, and will do so in such a way that all the predictions remain the same. That's the subject of the "renormalization group," so let's lay the groundwork for that.

NATURAL UNITS

Effective field theory might seem too good to be true. Certain parts of your theory (high-energy virtual particles) are causing you trouble? Just ignore them and stick to the parts that are working. It seems extraordinarily lucky, like turning your back to the basketball hoop and throwing up a shot over your shoulder that just happens to go in.

The reason why it works can be understood in terms of dimensional analysis. Remember that all physical quantities are measured in terms of certain systems of units: mass, energy, distance, time, and so on. Dimensional analysis is just the process of making sure that in any equations, the units (also called "dimensions," although they have nothing to do with directions in spacetime) of any two quantities that we add together or equate have to match. You can't add a number of centimeters to a number of grams.

Most physical units can be thought of as combinations, through multiplication or division, of a basic set of three: distance [D], time [T], and mass [M]. Consider something like energy. The Newtonian formula for kinetic energy is $E = \frac{1}{2}mv^2$. The pure number ½ is said to be "dimensionless." So, using square brackets to denote "dimensions of," we have

$$[\text{energy}] = [\text{mass}][\text{velocity}]^2 = [M][D]^2/[T]^2, \qquad (6.4)$$

since velocity has units of distance divided by time.

We have been using what particle physicists call **natural units**, in which

$$\hbar = c = 1. \qquad (6.5)$$

EFFECTIVE FIELD THEORY

This immediately implies relationships between the different basic units. For the speed of light we have

$$[c] = [D]/[T] = 1. \tag{6.6}$$

Consequently, when $c = 1$ we can say that distance and time are measured in the same units, $[D] = [T]$. You can measure time in centimeters if you like. To convert back to normal units, just multiply or divide by factors of c until you get the units you want. Likewise, for the reduced Planck constant $\hbar = 1.05 \times 10^{-27}$ g cm^2 s^{-1} we have

$$[\hbar] = [M][D]^2/[T] = 1. \tag{6.7}$$

This is a little harder to quickly put into action, but the philosophy is the same.

When we combine $c = 1$ with $\hbar = 1$, we find that all three of the basic units of mass/distance/time can be reduced to a single common scale. Particle physicists usually choose that scale to be energy, although they use "mass" and "energy" interchangeably. We already made use of this above when we expressed particle masses in terms of electron volts, which are really units of energy. Common abbreviations include GeV for giga (10^9) eV, MeV for mega (10^6) eV, and meV for milli (10^{-3}) eV. You will also run across keV for kilo (10^3) eV, TeV for tera (10^{12}) eV, and μeV for micro (10^{-6}) eV.

Appropriate application of (6.6) and (6.7) reveals that we can express all the basic units in terms of energy as

$$[M] = [E], \quad [D] = 1/[E], \quad [T] = 1/[E]. \tag{6.8}$$

So the units of every quantity can be thought of as energy to some exponent. Distance and time are just inverse energies, consistent with the physical relationship (6.3). (When $c = 1$, momentum and energy have the same units.)

DIMENSIONAL ANALYSIS

Now we can actually perform some of the dimensional analysis promised above, starting by thinking about Lagrangians and figuring out the dimensions of everything in units of energy to an appropriate power. Our original definition of the Lagrangian L was kinetic energy minus potential energy, so clearly the Lagrangian itself has dimensions of [E]. But in field theory we more often use the Lagrange density, related to the Lagrangian by

$$L = \int \mathcal{L}\, d^3x. \tag{6.9}$$

(We are now explicitly denoting all three dimensions of space, since the number of spatial dimensions will affect the units of things.) The notation d^3x refers to an infinitesimal volume element of space: length times width times height. It therefore has units

$$[d^3x] = [D]^3 = 1/[E]^3. \tag{6.10}$$

The integral sign has no dimensions. So to get the Lagrangian, with units of [E], we multiply the Lagrange density by the volume element, which has units $1/[E]^3$. Therefore, the Lagrange density itself must have units

$$[\mathcal{L}] = [E]^4. \tag{6.11}$$

Consider a simple scalar field theory, with Lagrange density (5.4), kinetic minus gradient minus potential energy densities. We want to figure out the units of the field ϕ. We can get there by considering

EFFECTIVE FIELD THEORY

any of the terms (they are added together, so must have the same units), so let's look at the kinetic energy density $\frac{1}{2}(\partial\phi/\partial t)^2$. A partial derivative with respect to time has units $1/[T] = [E]$ (because time is in the denominator of $\partial/\partial t$), and we want the whole term to have units of a Lagrange density, $[E]^4$. The field therefore has dimensions of energy,

$$[\phi] = [E]. \qquad (6.12)$$

This will hold for essentially all scalar fields, although other kinds of fields might have different dimensions. Electrons and other fermion fields, for example, have units $[E]^{3/2}$.

Those are the two basic facts we need: the scalar field has units of $[E]$, and the Lagrange density has units of $[E]^4$. When we write down interaction terms, they each come with a certain coupling constant, which tells us the strength of the corresponding vertex in a Feynman diagram. Every coupling constant will have to have units that make the interaction term as a whole have dimensions of $[E]^4$, because it's a term in the Lagrange density.

RENORMALIZABILITY

Why do we care so much about the units of the fields and the Lagrangian and its coupling constants? Because in an effective field theory, those coupling "constants" are not really constant, they depend on the value of the UV cutoff Λ. And that dependence will be just what we need to ensure that our effective theory gives us the same predictions, no matter what cutoff we choose to impose. Even better, it will imply that many interactions we could conceivably write down are ultimately irrelevant for making accurate predictions.

For simplicity's sake, consider a theory of a scalar field ϕ with an interaction Lagrange density consisting of three different terms:

$$\mathcal{L}_{int} = c_3\phi^3 + c_4\phi^4 + c_5\phi^5. \tag{6.13}$$

Remember that this is what working quantum field theorists spend a lot of time doing: starting with a free field theory, then writing down interaction Lagrangians and thinking about their consequences. To enter the particle physics mindset, we're going to stare at (6.13) for a while and think about what each term physically represents, and how they are affected by renormalization.

Each term has a physical interpretation as a certain number of particles interacting with each other, as manifested in vertices that we can stick together to make full Feynman diagrams. The simple terms in (6.13) lead to vertices with three, four, or five particles involved, one particle for each appearance of the field itself. Each term is associated with its own coupling constant.

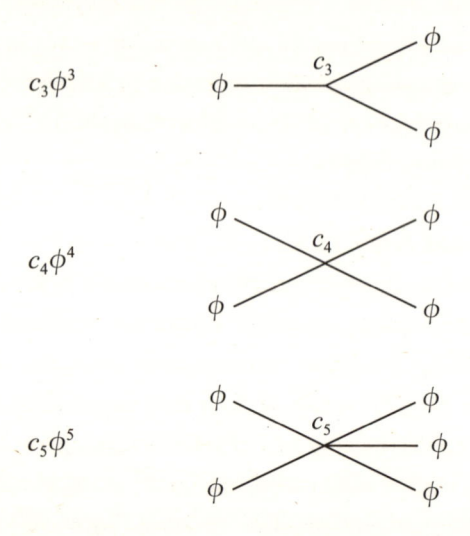

Our plain-vanilla scalar particles aren't distinguished from their own antiparticles, so it doesn't matter whether we draw the lines as incoming or outgoing; there is a corresponding vertex for every arrangement we can imagine.

EFFECTIVE FIELD THEORY

We need to think about the units associated with these coupling constants, which in turn will tell us how they are affected by our choice of our UV cutoff Λ. Because ϕ has units [E] and $\mathcal{L}_{int}$ has units of $[E]^4$, we immediately know the units of our couplings:

$$[c_3] = [E], \quad [c_4] = 1, \quad [c_5] = 1/[E]. \tag{6.14}$$

More generally, for a term of the form $c_n \phi^n$, the coupling will have units $[c_n] = [E]^{4-n}$. That 4 in the exponent is just the dimensionality of spacetime; in hypothetical theories in higher dimensions, coupling constants will have different units.

In the mid-twentieth century, when physicists were wrestling with infinities in QFT and before the EFT approach had been developed, they quickly realized that the renormalization procedure that worked for quantum electrodynamics wouldn't work for all theories. Some theories are **renormalizable**, but some are not. In a renormalizable theory, the infinities we get from loop diagrams can be canceled in terms of a small number of parameters, leaving us with a nice predictive theory. QED is a paradigmatic example.

And for the most part, the distinction is simple: in a renormalizable theory, all the coupling constants have units $[E]^a$ with $a \geq 0$. Their dimensions are energy to some non-negative power. In our simple scalar model, c_3 and c_4 would be allowed, but we would have to set $c_5 = 0$ for our theory to be renormalizable.

Roughly speaking, then, Lagrangian terms with small numbers of fields are renormalizable, while those with large numbers of fields are not. And there aren't that many terms you can write down with small numbers of fields. So the demand that a theory be renormalizable was really helpful to the working physicist: only a very restricted class of possible theories would qualify.

Once you include interaction terms with couplings whose dimensions are negative powers of energy—which happens in our current example when there are more than four appearances of the scalar field, like the $c_5\phi^5$ term—the theory is **non-renormalizable**. When your coupling constants have units of negative powers of energy, the resulting expression for the Feynman diagrams needs to have positive powers of the momentum, to make the overall dimensions come out right. So when we integrate up to large momentum, it's easy to get infinity. Indeed, there will generally be an infinite number of infinities, each of which requires special treatment—introduction of a new term to cancel each infinity in precisely the right way. We can't physically measure an infinite number of observables to pin down the features of all these new terms. So the problem with non-renormalizable theories isn't that they are fundamentally sick or ill-defined; it's that they require an infinite number of inputs in order to make any experimental predictions, which seems like a bad quality for a physical theory to have.

At the time, physicists responded to this situation by insisting that they should only consider renormalizable theories. That was enormously helpful, as it narrowed down the potentially infinite number of terms you could put into a Lagrangian to just a manageable handful. In the 1960s, when the idea of a Higgs boson was first proposed and it was conjectured that it might help unify the electromagnetic and weak nuclear forces, most physicists didn't pay close attention because they thought that the theory wouldn't be renormalizable. Then in the early 1970s, Gerard 't Hooft and Martinus Veltman showed that the electroweak theory was renormalizable after all. Suddenly everyone got caught up in the excitement and started taking the theory seriously. The new particles it predicted were all ultimately discovered, and 't Hooft and Veltman won the 1999 Nobel Prize in Physics for their work.

EFFECTIVE LAGRANGIANS
The effective field theory paradigm shifts this perspective a bit. Let's imagine you have our scalar field theory with interactions (6.13), but

EFFECTIVE FIELD THEORY

we set $c_5 = 0$ so the theory is renormalizable. Even without that fundamental five-particle vertex, there will still be processes involving five real particles, with (for example) two incoming particles and three outgoing ones. For example, we can combine three-particle vertices into a one-loop diagram with five external particles overall, as shown in the figure.

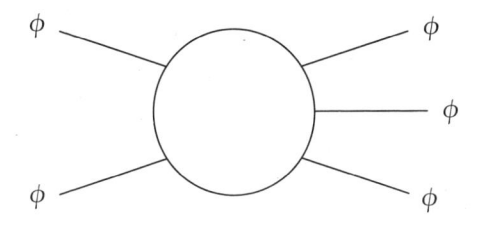

From the EFT perspective, this gives rise to an *effective* five-particle interaction. The rules of the game are that we construct an entire **effective Lagrangian**, where we start from the fundamental (renormalizable) Lagrangian and then rather than doing the old-fashioned thing where we integrate loop momenta all the way up and subtract off infinities, instead we adjust the values of the coupling constants to include all of the effects that would have been induced by them. So it isn't accurate to imply that arbitrarily large loop momenta are simply thrown away; we say they are **integrated out**, and their effects show up in the effective coupling constants. We are bundling up unknown UV processes into a simple set of IR parameters.

The effective Lagrangian doesn't just include the renormalizable terms we started with; it includes every term we can think of. So it will have a term of the form $c_5 \phi^5$, and indeed all possible terms $c_n \phi^n$ with higher n's, renormalizability be damned. But the apparent non-renormalizability doesn't bother us, because we don't use that Lagrangian to then integrate loop momenta up to infinity; we integrate only up to the UV cutoff Λ, and always get finite answers.

This procedure helps answer a question that may have been eating at you for a while. We invented an effective field theory by introducing an ultraviolet cutoff Λ, then not including UV momenta above that scale when we did loop integrals in our Feynman diagrams. But how do we choose what Λ should be? If that choice affects what contributions we have in our diagrams, it sounds like it will have a big effect on what our theory predicts.

That supposition is wrong. The choice of Λ has no effect at all on our theoretical predictions. The UV virtual particles can have an effect, but all of those effects are bundled up in the values of the coupling constants in the effective Lagrangian. That's how we end up with a theory that only needs to talk about IR particles.

To put some meat on these bones: the mass of the photon is 0 and the mass of the electron is 0.511 MeV (million electron volts). Let's say we set our ultraviolet cutoff to 1 keV (thousand electron volts), well below the electron mass. Then the electron is a UV particle, and we have a theory with nothing but photons in it. The Lagrangian for QED doesn't include any direct couplings between photons and other photons, only between photons and electron/positron pairs. But we can write down an EFT with a 1 keV cutoff that includes photons interacting with each other; those non-renormalizable interactions are induced by electron loops at higher energies. (Heisenberg and his colleague Hans Heinrich Euler actually calculated this effect back in 1936, without the benefit of Feynman diagrams or effective field theory.) This effective four-photon interaction appears as a basic vertex in the EFT, summarizing a whole bunch of UV processes that we're not explicitly calculating.

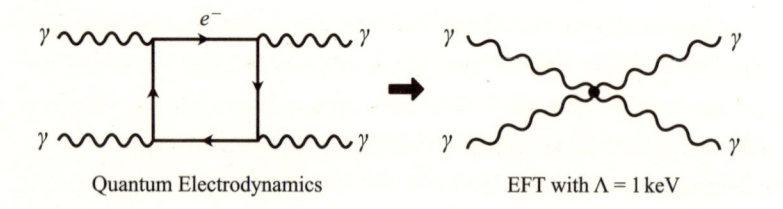

Quantum Electrodynamics   EFT with $\Lambda = 1$ keV

EFFECTIVE FIELD THEORY

THE RENORMALIZATION GROUP

To make all this work, we need to recognize that the coupling "constants" appearing in the EFT Lagrangian aren't really constant at all. They are functions of Λ. That makes sense. The meaning of a coupling constant in the Lagrange density is simply the "strength" of the corresponding interaction. The fine-structure constant α, for example, tells us the strength of the electromagnetic interaction. And in an effective field theory with cutoff Λ, we are bundling up all of the effects of virtual particles with energies *above* Λ into the value of the effective coupling. So of course the effective coupling will depend on the value of Λ, since that governs what the UV processes are that we are bundling up.

We therefore say that the coupling constants "flow" or "run" as a function of the cutoff, and refer to them as **running coupling constants**. Instead of simply writing c_4 as the coefficient of ϕ^4 in our scalar field theory, we would write it as a function of the cutoff, $c_4(\Lambda)$. The way the coupling constants change for different values of the cutoff is called the **renormalization group**.

In fact, we can make a decent guess as to how our coupling constants will evolve as we change Λ, just using dimensional analysis. We saw back in (6.14) how different couplings have different units, and we know that the effective couplings are functions of Λ. So the simplest idea, which turns out to be pretty accurate, is that the couplings are proportional to appropriate powers of Λ:

$$c_3(\Lambda) \propto \Lambda, \quad c_4(\Lambda) \propto \Lambda^0, \quad c_5(\Lambda) \propto \frac{1}{\Lambda}, \ldots \qquad (6.15)$$

We see an obvious pattern. If our interaction ϕ^n has $n < 4$, then its coupling constant is going to be proportional to a positive power of the cutoff Λ. As we go to bigger and bigger values of Λ, that coupling will become larger and larger. We refer to such interaction terms as **relevant operators**, since fields and combinations thereof can be thought of as operators on the Hilbert space of the quantum field theory. When we have interactions involving more fields, ϕ^n with $n > 4$, the coupling will go as a negative power of Λ, and we call those **irrelevant operators**. They get smaller as Λ grows.

For ϕ^n with $n = 4$, the coupling is dimensionless, and such terms are dubbed **marginal operators**. A marginal operator can evolve with Λ, but the running will be more gradual—typically as a logarithm of Λ—than the simple power law we get for relevant and irrelevant operators. It's not obvious at first glance whether a marginal operator will grow or shrink as we increase the cutoff; you have to do an explicit calculation to find out. In quantum electrodynamics, the fine-structure constant α is a dimensionless coupling that grows at high energies. In fact, it becomes infinitely large at a super-high scale known as the **Landau pole**, named after Soviet physicist Lev Landau. That's a problem, and it is one reason why many people think QED needs to ultimately be subsumed into a better-behaved unified theory. By contrast, in quantum chromodynamics (QCD), which is the theory of the strong nuclear force, the coupling diminishes at high energies. This is known as **asymptotic freedom**, since quarks and gluons become free (non-interacting) as the energies go up. QCD seems to be a perfectly well-behaved quantum field theory, up to arbitrarily high energies.

The designations relevant/marginal/irrelevant for operators (terms in the Lagrangian) all make sense. The choice of UV cutoff is up to us, with one rule: it has to be at a higher energy than any of the particle masses or momenta that we are considering at the moment. That is, Λ must be large compared to the scales of interest in our problem. As we consider larger and larger values, the couplings corresponding to relevant operators will grow, those for irrelevant operators will shrink, and those for marginal operators won't change that much. So the irrelevant operators can often, at least at a first pass, be neglected.

Which explains why renormalizable theories seemed worthy of

special attention in the first place. We don't know what the ultimate theory of everything is, or even whether it's a quantum field theory at all, certainly not down to ultra-small scales and high energies. But an effective field theory that derives from that ultimate theory will generally *look* renormalizable, at least to a good approximation. The effects of UV loop momenta tend to make the renormalizable (relevant and marginal) terms look big, and the non-renormalizable (irrelevant) terms look small. All manner of crazy things might be going on at tiny distances, and the world will still look nice and renormalizable to us.

The list below summarizes the situation. (We're including only terms with an even number of fields, to make things simple.) In old-fashioned quantum field theory, we would restrict ourselves to only renormalizable terms, integrate loop momenta up to arbitrarily high scales, and subtract off the resulting infinities by renormalization. In a modern effective-field-theory approach, we allow every term we can write down, but we only integrate loop momenta up to a UV cutoff Λ, which keeps everything finite, and the couplings depend on Λ so as to keep the physical predictions independent of Λ. Terms with many fields generally have irrelevant couplings that get tiny as Λ gets large, so we don't have to worry about them much anyway.

- Renormalizable scalar field theory:

$$\mathcal{L} = \frac{1}{2}(\partial_t \phi)^2 - \frac{1}{2}(\partial_x \phi)^2 - \frac{1}{2}m^2\phi^2 + c_4\phi^4$$

  - virtual particles have arbitrarily high energies
  - subtract infinities via renormalization

- Effective scalar field theory:

$$\mathcal{L} = \frac{1}{2}(\partial_t \phi)^2 - \frac{1}{2}(\partial_x \phi)^2 - \frac{1}{2}m^2(\lambda)\phi^2 + c_4(\lambda)\phi^4 + c_6(\lambda)\phi^6 + \cdots$$

- virtual particles have energies below UV cutoff Λ
- coupling constants depend on Λ so that physical predictions do not

One annoying detail worth mentioning: it might seem puzzling that something as direct and measurable as the mass of the particle could depend on the cutoff. You are right to be puzzled. The resolution is that in this picture we lose the nice equivalence between the particle's mass and the parameter m that appears in the Lagrangian. There is still a "physical" particle mass; it's not equal to m, but we can calculate it as a combination of $m(\Lambda)$ and the higher-order couplings like $c_4(\Lambda)$ and so on. Physicists have reconciled themselves to the fact that some things are just complicated.

MYSTERIES OF NATURALNESS

The effective field theory paradigm is both intellectually compelling and practically useful. It plays a central role in modern particle physics and has been extended to contexts as diverse as the growth of structure in cosmology and the generation of gravitational waves by inspiraling black holes.

But there are looming worries. These center on the notion of the **naturalness** of parameters in particle physics. Naturalness is not a rigorously formulated principle, but it provides motivation for much research in contemporary particle theory. The basic idea is that numerical parameters in an effective Lagrangian should not be very small compared to 1, unless there is a symmetry or other good reason for such a small number to appear.

What we have in mind is that, thinking about Feynman diagrams, the low-energy effective parameters we measure are actually the sum of many different contributions. Some contributions could be negative as well as positive. It would be natural to expect the sum of all these contributions to be of the same order as the individual contributions. In other words, if we added up a bunch of ten-digit numbers (some positive, some negative), it would be strange if the result were exactly zero, or even any number between −10 and +10. That could conceivably happen, but it would appear to be an unusual coincidence of some sort, often referred to as **fine-tuning**. (Nothing to do with fine-tuning of parameters to allow for the existence of life, which you also hear discussed in cosmological/theological discussions.)

And yet, that apparently does happen in the real world, in two famous cases. One is the mass of the Higgs boson, $m_H = 125$ GeV (giga electron volts = 1 billion eV). That might seem like a big number, over a hundred times the mass of the proton. But if we label the Higgs field H, the mass term in the effective Lagrangian is simply $\frac{1}{2} m_H^2 H^2$. It's a relevant operator, and by dimensional analysis we would expect the coupling m_H^2 to go as Λ^2. Admittedly that's a bit strange, since the UV cutoff Λ is supposed to represent a parameter we introduce for our own convenience, not a true constant of nature. But in this context it simply reminds us that loop diagrams work to increase the effective value of the Higgs mass, as far as they can. New physics might kick in at some high-energy scale like the Planck scale, $M_P = 10^{18}$ GeV. But then we would expect the measured Higgs mass to be near the Planck scale, and it's not. They differ by quite a bit:

$$\frac{m_H}{M_P} \sim 10^{-16}. \tag{6.16}$$

That's exactly the kind of small number the naturalness principle would seem to forbid, but there it is. This discrepancy between expectation and reality is known as the **hierarchy problem** in particle physics. It could be solved if, for example, there were new physical effects near an energy scale of 100 GeV that worked to stabilize the

Higgs mass. One example is supersymmetry—a hypothetical symmetry between bosons and fermions—that could act to cancel different loop contributions against each other. There have been other proposals, but they generally all predict that we should see new particles near this energy scale. The search for such particles was a major motivation for building the Large Hadron Collider, but as of this writing no such particles have been found. The hierarchy problem remains unsolved. (Other particles have masses, of course, but in every known case other than the Higgs there is a symmetry or other mechanism that explains why their masses are so small compared to the Planck scale.)

A similar problem is the **cosmological constant problem**. As briefly noted in *Space, Time, and Motion*, the cosmological constant was introduced by Einstein as a modification of his equation for general relativity. From the perspective of field theory, it can be thought of as the simplest possible term we could include in an effective Lagrangian: a bare constant ρ_0, not multiplying any fields at all. We can interpret ρ_0 as representing the energy density of empty space—the **vacuum energy**. The vacuum energy and the cosmological constant are two different labels for exactly the same thing; don't let anyone tell you any different.

By dimensional analysis, we have $[\rho_0] = [E]^4$. From an EFT perspective with cutoff Λ, we would expect it to be of order $\rho_0 \sim \Lambda^4$. If we have no novel physical processes that could kick in to cancel out contributions from loop diagrams, we might expect to plug in the Planck mass M_p once again as a reasonable value for Λ here. This time our expectations are even more dramatically off than in the case of the hierarchy problem. Observationally we have

$$\frac{\rho_0}{M_P^4} \sim 10^{-122}. \qquad (6.17)$$

That's a big discrepancy, famously the largest mismatch between expectation and observation in all of theoretical physics.

The hierarchy problem and the cosmological constant problem are both, at face value, pretty spectacular failures of our EFT intuition. To date there is no consensus on how to address them. There is no shortage of proposals, but none of them has really caught on. We don't need to throw out the whole EFT program, but perhaps there are subtle means by which UV effects are reaching out to tune our IR parameters in unanticipated ways. Or maybe even something more dramatic is going on. It's when our theories don't quite agree with our data that science gets exciting.

SEVEN

SCALE

The evocative film *Powers of Ten* by Charles and Ray Eames (based on an earlier book by Kees Boeke) invites the viewer to experience the scope of the universe by starting with a single square meter view—a couple enjoying a picnic lunch on Chicago's lakefront—and zooming out by a factor of ten every ten seconds. We see a gradually shrinking city, the Earth, the solar system, nearby stars, the Milky Way galaxy, then clusters of galaxies and even larger structures. We then zoom back in and start another journey deeper into scale, seeing skin cells, organelles, molecules, atoms, and elementary particles.

There are substances, such as dark matter or background neutrinos, that we don't directly see in such images. Everything we do see is either an atom or made of atoms. And atoms are made of just three constituents: protons, neutrons, and electrons. We sometimes take for granted the enormous hierarchy of scale that makes up the journey from elementary particles to the universe.

Our last few chapters were a challenging survey of deep theoretical ideas. In this chapter we'll get a bit of a breather by considering a topic that is easier to visualize: what the known universe is actually made of, and in particular the various scales of mass and energy that characterize our world.

UNITS REVISITED

We've been using "natural units," in which $\hbar = c = 1$. Then we have the following relationships between dimensions of energy, mass, distance, and time:

$$[E] = [M] = 1/[D] = 1/[T]. \tag{7.1}$$

Having made this choice, we still need to pick a fundamental unit to use, and particle physicists most often pick energy. In particular they will measure energy in **electron volts** (eV), defined as the kinetic energy that would be gained by an electron as it accelerates through one volt of electrical potential. Most people, including professional particle physicists, don't have much physical intuition for pushing individual electrons across voltages. But other energy units aren't much more intuitive. Household energy use is often measured in kilowatt-hours (kWh) or British thermal units (BTU). Energy stored in food is typically measured in calories, although the numbers you find in dietary guidelines or nutrition listings are actually kilocalories (kcal). In the metric system we can either use joules (equal to kg m^2/s^2) or ergs (g cm^2/s^2). The following relations let you convert back and forth.

$$\begin{aligned}1 \text{ eV} &= 1.60 \times 10^{-19} \text{ J} = 1.60 \times 10^{-12} \text{ erg} = 4.45 \times 10^{-26} \text{ kWh} \\ &= 1.52 \times 10^{-22} \text{ BTU} = 3.83 \times 10^{-23} \text{ kcal.}\end{aligned} \tag{7.2}$$

We see at least one convenient feature of the electron volt: it's a tiny amount of energy. Particles are tiny things, so these kinds of units make sense.

SCALE

Thinking like a particle physicist, we can convert common units to and from electron volts.

$$1 \text{ eV} \approx 10^{-33} \text{ grams},$$
$$1 \text{ eV}^{-1} \approx 10^{-5} \text{ cm},$$
$$1 \text{ eV}^{-1} \approx 10^{-15} \text{ sec}. \quad (7.3)$$

We shouldn't interpret these relations as implying that we can physically transform an amount of distance into an amount of energy, or anything like that. They are simply choices of units with which to measure things. But as we will see, they provide rough rules of thumb to relate common distances and times to particle physics processes. One electron volt is a small amount of mass, and its inverse is a small distance and a short period of time, at least by human standards.

THE SCALES OF PARTICLE PHYSICS

Almost no particle physicist could tell you the mass of the electron in grams. But they would almost certainly be able to recite its mass in eV, along with that of other familiar particles.

Electron	5.11×10^5 eV	0.511 MeV
Proton	9.383×10^8 eV	0.9383 GeV
Neutron	9.396×10^8 eV	0.9396 GeV
Neutrino	$\leq 10^{-1}$ eV	≤ 100 meV

One GeV is a convenient unit for use in particle physics, since it's approximately the mass of a proton or neutron. We've included here the neutrino, which we've mentioned but haven't really talked about in detail. Neutrinos are low-mass, neutral, nearly invisible particles that turn out to have remarkably subtle properties.

To get our bearings, it's helpful to portray these and other particle physics scales on a logarithmic scale. We've included not only the masses of the above particles but also the reduced Planck mass ($\sqrt{\hbar c / 8\pi G}$, where G is Newton's gravitational constant), typical collision energies at the Large Hadron Collider, the mass of the Higgs boson, atomic nuclei, and the vacuum energy scale E_{vac}. To define the latter, we note that the vacuum energy density ρ_0 has units $[E]^4$, and define $E_{vac} = \rho_0^{1/4}$. Finally, we included the hypothetical energy scale of **Grand Unification**, a scenario that would unify the strong, weak, and electromagnetic forces (but not gravity).

- 10^{18} GeV ← Planck scale / moving train
- 10^{15} GeV ← Grand Unification?
- 10^{12} GeV ← hit baseball
- 10^{9} GeV
- 10^{6} GeV ← flying mosquito
- TeV ← Large Hadron Collider
- — Higgs
- GeV ← nuclei / protons, neutrons
- MeV ← electrons
- keV
- eV
- meV ← neutrinos / vacuum energy

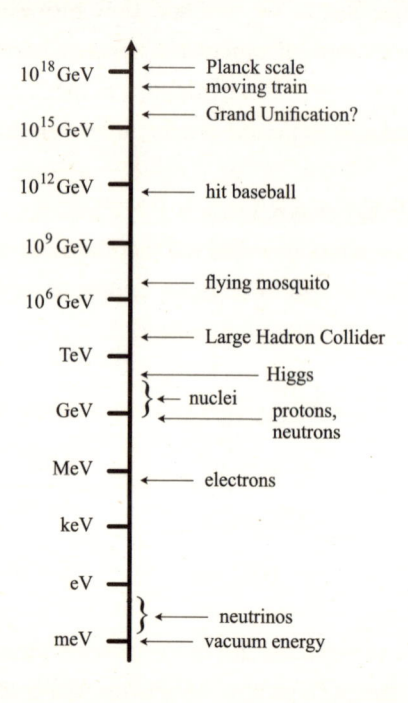

For comparison we've included the kinetic energies of some everyday phenomena: a flying mosquito, a hit baseball, and a fast-moving train. (For the listed particles we are quoting rest energies—that is, masses, not kinetic energies.) These everyday energies are generally higher than the energies at the Large Hadron Collider or other high-energy accelerators, which might come as a surprise. Why do we have to spend so much money on a giant machine that can't even match the energy of a mosquito? The answer, of course, is that the collision energy at the LHC is concentrated into just two protons. Mosquitoes and other macroscopic objects have more energy, but only because they are made of many particles; their energy per particle is relatively low.

When scientists say "many particles," it's useful to have a rough order of magnitude in mind. A convenient choice is **Avogadro's number**, $N_A = 6.02 \times 10^{23}$. That's roughly the number of **nucleons** (protons and neutrons) in one gram of atomic matter. (When it comes to atoms, electrons are almost irrelevant to their mass.) Classical macroscopic behavior can kick in for much smaller quantities, but Avogadro's number provides a good starting point for connecting atoms to human-scale objects. One gram is fairly small—about the mass of a paperclip. Here are some other familiar scales for comparison:

- Bacterium or human DNA $\sim 10^{12}$ nucleons
- Person $\sim 10^{28}$ nucleons
- Earth $\sim 10^{51}$ nucleons
- Sun $\sim 10^{57}$ nucleons
- Milky Way galaxy $\sim 10^{70}$ nucleons
- Observable universe $\sim 10^{80}$ nucleons

It's interesting to contemplate these numbers for their own sake. Our universe is a thicket of hierarchies, with coherent structures of vastly different sizes. For this chapter we will content ourselves with thinking about the sizes of individual atoms.

THE SIZE OF A QUANTUM PARTICLE

With these orders of magnitude in mind, we can think about connecting them back to the world of quantum mechanics and quantum field theory. Our route into doing that will be to relate mass and distance via a crucial concept: the **Compton wavelength** of a particle. It represents the minimum amount of space a single particle can be said to occupy.

When Louis de Broglie first suggested that particles had wave-like properties back in 1924, he proposed the relevant wavelength to have in mind for a particle with momentum p, what we now call the de Broglie wavelength,

$$\lambda_{dB} = 2\pi/p. \qquad (7.4)$$

(We wrote it earlier as h/p, but now we're setting $\hbar = h/2\pi = 1$.) This is the length scale we associate with the wave-like behavior of the quantum wave function of a massive particle. When we calculate the distance between interference bands in the double-slit experiment, or require that an atomic orbit contain an integer number of wavelengths, it's the de Broglie wavelength we have in mind.

The Compton wavelength plays a different role. It is defined for a particle of mass m by

$$\lambda_C = 1/m. \qquad (7.5)$$

(This is the "reduced" Compton wavelength; the "regular" one is this times 2π.) Unlike the de Broglie wavelength, the Compton wavelength has nothing to do with the motion of the particle; it's a fixed constant, once we know the mass. The Compton wavelength for an electron is approximately 2×10^{-10} cm, while for a proton it is approximately 1×10^{-13} cm. The notion was introduced by Arthur Compton

SCALE

in 1923 in his investigation of photon-electron scattering. But it really comes into its own in quantum field theory, where it expresses a difficulty in pinning down the spatial extent of a particle.

According to relativity, the energy is related to the momentum and mass of a particle by

$$E^2 = p^2 + m^2. \tag{7.6}$$

So a particle at rest, with $p = 0$, has $E = m$, which we recognize as a famous equation once we remember we're setting $c = 1$. But we also have the Heisenberg uncertainty principle, according to which the spread in the position and momentum expressions for a particle wave function must satisfy

$$\Delta x \cdot \Delta p \geq 1. \tag{7.7}$$

Imagine that we have a particle whose wave function is spatially localized in a region the size of its Compton wavelength, so that

$$\Delta x \sim 1/m. \tag{7.8}$$

(Remember that $\sim$ means "of order," so in relations like this we are ignoring numerical factors of order unity—that is, between 0.1 and 10.) Then we can't just set the momentum to 0, since there's going to be some uncertainty for it, of order

$$\Delta p \sim m. \tag{7.9}$$

Because of this momentum uncertainty, from (7.6) there will also be some uncertainty in the energy, even if the average value of the momentum is $p = 0$:

$$\Delta E \sim \Delta p \sim m. \qquad (7.10)$$

So if we try to localize a particle to within its Compton wavelength, the energy uncertainty has magnitude $\Delta E \sim m$. But that's enough energy to make an entire additional particle, a process that QFT is all too happy to accommodate. And notice that if we were to localize the particle to an even smaller region, the energy uncertainty would be even larger.

Within the framework of quantum field theory, this means that if you try to squeeze a particle into a region smaller than $\lambda_C = 1/m$, the quantum state becomes a superposition of one or more particles—not a single particle at all. The Compton wavelength is the smallest length scale at which we can safely think about the system as just representing one particle. In practice, this means that the Compton wavelength can be thought of as the "size" of a particle, as far as QFT is concerned. And notice that it goes down as the mass goes up. In quantum field theory, heavier particles occupy less space. It's a funny concept of size, however, not completely harmonious with our classical intuition. The wave function of a particle can easily be spread out over a distance greater than its Compton wavelength; it just can't be squeezed into less.

There is an immediate and sad implication of this fact: Ant-Man, the superhero who can shrink down to arbitrarily small sizes, is not really possible. Nor can we imagine that there are entire organisms or civilizations within a single atom. If you tried to shrink Avogadro's number of particles down to the size of a proton, either each of those particles would have to be more massive than the proton itself—in which case we would notice, because the thing would be much heavier—or they would be lower-mass, and their Compton wavelengths would be too large to fit in such a small space. For all of its surprises, the microscopic quantum realm is relatively smooth and featureless. Small wiggles carry too much energy.

QUARKS, HADRONS, BARYONS, MESONS

The idea that we can't squeeze a collection of many particles down to subatomic scales might seem to bump against another fact you may have heard: every proton and neutron is made of three **quarks**. In the 1960s Murray Gell-Mann and (independently) George Zweig realized that one could make sense of the rapidly accumulating set of strongly interacting particles being discovered in particle accelerators by positing that they were all made of smaller constituents, dubbed quarks by Gell-Mann after a line in James Joyce's *Ulysses*: "Three quarks for Muster Mark!" Inspiration for the name came from the idea that protons and neutrons, and many of their hadronic cousins, contained three quarks each. Presumably the quarks have less mass than the protons and neutrons that they combine to form, and therefore larger Compton wavelengths. So how can they be squeezed inside? This is as good a time as any to talk a bit about quarks and the particles they form, called **hadrons**.

Yes, protons and neutrons are made of quarks, and also of the massless bosons carrying the strong nuclear force that holds them together, called **gluons**. There are six known **flavors** of quarks—**up, down, charm, strange, top, and bottom**—but only the lightest examples, the ups and downs, contribute substantially to protons and neutrons. (Heavier quarks play a small role as virtual particles.) The up quark has electric charge $+2/3$ and the down quark has charge $-1/3$, and there are three quarks in each nucleon. From playing around a bit with those charges, you can convince yourself that there must be two ups and a down in each proton, and two downs and one up in each neutron. You'd be correct. Fermions that do not feel the strong interactions are known as **leptons**, of which there are six flavors known: the electron, muon, and tau, all with charge -1, and corresponding neutrinos, the electron neutrino, muon neutrino, and tau neutrino, all with charge 0.

The mass of the up quark is about 2.2 MeV, and the mass of the down quark is about 4.7 MeV. We immediately notice two things: First, the combined mass of the quarks in a proton is much less than the proton mass itself (938.3 MeV), and likewise for the neutron (939.6 MeV). Where does all the extra mass come from? Second, the Compton wavelength of each quark is indeed much bigger than the Compton wavelength of the nucleons themselves. How do they fit?

Both questions arise because the interior of a nucleon is a place where our intuition, trained as it is on particles and classical mechanics, fails us utterly. Even our fancy Feynman diagram ways of thinking aren't going to be up to the task. The interactions between quarks and gluons are strong, not weak. We found that QFT predicts particles by looking first at a free field theory (no interactions at all) and then introducing interactions between the particles using perturbation theory. Inside a nucleon there is no small parameter like the fine-structure constant to make perturbation theory possible, and we can't safely think about what's going on in terms of a collection of weakly interacting particles. Nucleons are thoroughgoingly quantum-field-theoretic objects, and we need to take QFT seriously to understand them.

The fact that the strong interactions are indeed strong leads to the crucial feature that helps answer our questions: **confinement**. Electric charge is a single variable that can be positive, negative, or zero. At the risk of sounding unnecessarily recondite, think of electric charge as taking values in a one-dimensional vector space. According to the theory of **quantum chromodynamics** (QCD), quarks have an additional kind of **color charge** that takes values in a *three*-dimensional vector space. This, of course, has absolutely nothing to do with "color" as we typically perceive it with our eyes, it's just a whimsical name.*
But it serves a useful purpose. If we label the three axes of color-charge space as "red," "green," and "blue," the principle of confinement says

---

* Quark "flavors" aren't flavors you can taste, either. Sorry to disappoint.

that only colorless ("white") combinations make observable particles. Quarks are individually going to be red, green, or blue, and we will never see them alone in the wild. What we see are combinations of three quarks, each of a different primary color, adding together to give a colorless proton or neutron.

Think of it this way. A single hydrogen atom contains one electron and one proton. But its mass is slightly less than the sum of those two particles, by about 13.6 eV. That's because both the electron and the proton, by themselves, are surrounded by electric fields, and these fields carry energy. In an atom, those fields (largely) cancel each other out, so that energy isn't there. As a result, the total energy (and therefore mass) is a bit smaller. The 13.6 eV discrepancy is the **binding energy** of the electron in hydrogen. You have to put at least that much energy in to take the two particles apart.

Now consider a quark all by itself, not inside a proton. There is no such thing, but imagine that there were. It would be surrounded by a chromodynamic field, the field that gives rise to gluons in the same way an electromagnetic field gives rise to photons. But whereas the electromagnetic field falls off as an inverse square (Coulomb's law), the chromodynamic field gets bigger and bigger with distance from the quark. Consequently, there would have to be an infinite amount of energy in that gluon field. Whereas if you take three quarks in a colorless combination, the gluon field cancels out beyond a certain distance, just as the electric field does in an atom. You're left with a finite-energy nucleon, in which the binding energies of the quarks are effectively infinite.

There are other colorless combinations of three quarks besides protons and neutrons: particles that involve different quark flavors, or higher-energy states of the three quarks. Collectively such particles are known as **baryons**. As far as current experiments have revealed, the total number of baryons minus the total number of anti-baryons in a process—the net **baryon number**—is conserved. You can equally well think of that as conservation of the net "quark number," but we knew about conservation of baryon number before we knew baryons were made of quarks, so that's the label we use. The total number of baryons in the entire universe, however, is larger than the number of anti-baryons. This matter/antimatter asymmetry hasn't yet been explained by known physics. (**Lepton number**, the number of leptons minus anti-leptons, is separately conserved. But we don't know whether there is an excess of leptons over anti-leptons, since we don't know the total number of neutrinos and anti-neutrinos in the universe.)

There is one more way to make a colorless combination: one quark and one anti-quark. Anti-quarks are either anti-red, anti-blue, or anti-green, which can be thought of as cyan, yellow, and magenta. A quark/anti-quark combination is called a **meson**, in contrast with the three-quark baryons. You might expect mesons to decay away pretty quickly, which they generally do, but depending on what kinds of quarks they're made of they can hang around long enough to be detected.

Now we're in a position to answer our two questions. Why are protons and neutrons so much more massive than their constituent quarks, and how do those low-mass quarks fit inside the Compton wavelength of a nucleon? The answers come down to the fact that thinking of a nucleon as simply three quark particles orbiting each other is a wildly misleading image. They are actually complicated quantum configurations of quark and gluon fields. The honest statement is just that the proton is the lowest-energy state with baryon number equal to 1. The total mass of a nucleon comes from the gluon fields even more than the quark fields; if quarks had zero mass, the masses of protons and neutrons wouldn't be appreciably different. And it shouldn't be surprising that the quarks can be squeezed into a region smaller than their Compton wavelength, even though that wavelength represents the smallest size for a single particle all by itself. The quarks inside baryons are very much *not* by themselves; they have settled into a particular quantum state in constant interaction with other quark fields and with the gluons.

You will sometimes hear descriptions of the inside of a proton as a wild place, with fluctuating gluons and quark/anti-quark pairs popping in and out of existence. Even professional physicists talk like this. By now we've learned enough quantum field theory to know better. Such a picture is just as misleading as thinking of an electron orbiting around a nucleus like a tiny planet in a solar system. The reality is a quantum wave function. The wave function of the fields in a proton is a specific, highly entangled combination of quark and gluon fields. It's not even precisely right, although in practice it's forgivable, to say "there are three quarks in a proton." There are three quarks that contribute to a proton's baryon number—known as the **valence quarks**—but the configuration is fundamentally not particle-like. And other fields do contribute to what's going on, much like virtual particles in a Feynman diagram.

The field configuration in a nucleon is also completely static—the wave function isn't changing from moment to moment in any way. If we were somehow able to perform a quantum measurement of the state of a quark or a gluon inside a nucleon, it would "collapse" to some particular location, and repeated measurements would collapse to different states, giving the impression of "random fluctuations." But for most protons, nobody is measuring inside them, and nothing is fluctuating at all. They are static solutions to the Schrödinger equation for the Standard Model with baryon number equal to 1.

BOHR RADIUS

That was intense. Let's take a leap upward in scale to the friendly confines of the hydrogen atom.

A basic hydrogen atom consists of a proton and an electron. In everyday units, those particles have (reduced) Compton wavelengths

$$\lambda_p = 2.1 \times 10^{-14} \text{ cm}, \quad \lambda_e = 3.9 \times 10^{-11} \text{ cm}. \tag{7.11}$$

The electron's wavelength is obviously bigger, so it's the electron that sets the size of the atom itself; the proton just sits localized at the center. This is why electrons do all the work in atomic and molecular physics, not to mention chemistry and biology.

Now let's look up the size of an actual hydrogen atom. That's not a completely well-defined figure, since we're talking about a smooth quantum wave function, not a solid object with a precise boundary. But a good approximate measure is given by the **Bohr radius**, defined as the most probable distance from the proton we might observe an electron to be, were we to measure it. It's given by

$$a_0 = 5.3 \times 10^{-9} \text{ cm}. \tag{7.12}$$

That is clearly not equal to the electron's Compton wavelength; it's over a hundred times bigger. What's going on?

The Compton wavelength is the minimum size at which a single particle can be localized, but it's certainly possible to spread that particle's wave function over a larger distance. In the case of the hydrogen atom, we have to consider the electrical force from the proton that is keeping the electron bound to it. We can think of this in terms of a potential energy as a function of distance:

$$V(r) = -\frac{\alpha}{r}, \tag{7.13}$$

where $\alpha \approx 1/137$ is the fine-structure constant. It's the fine-structure constant that tells us the strength of electromagnetism; the fact that it is small, but not too small, indicates that the electric binding force is weak, but not too weak.

What happens, then, is that the electron gets to spread out a little in the hydrogen atom, because it is being bound by the weak-but-not-too-weak electromagnetic potential. If we plug in numbers, we find

$$a_0 = \frac{\lambda_e}{\alpha}. \tag{7.14}$$

This should be the least surprising thing in the world, because it scales in the right way. The length scale for an electron in a hydrogen atom is indeed "set" by the electron's Compton wavelength, because that is the relevant parameter that has dimensions of length. But it is modified by the dimensionless parameter α, which characterizes the strength of the electromagnetic interaction. And it makes sense that it is inversely proportional to α: if we imagine taking $\alpha \to 0$, making electromagnetism weaker, the electron would become less and less bound to the proton, eventually spreading out over all of space. Whereas if we made α larger, the electron would be pulled closer to the proton.

This is a simple example of a **scaling argument**: starting with basic length/time/energy scales (like the electron Compton wavelength) and using physical intuition to figure out how they are affected by other parameters (like the fine-structure constant) to estimate some quantity of interest (like the Bohr radius). Note that we were able to figure this out without ever actually solving the Schrödinger equation in any detail, just using scaling and dimensional analysis. This is why physicists will always first think about those kinds of arguments before doing any hard work. (As the physicist John Wheeler once advised, "Never make a calculation until you know the answer.")

We can even go a bit further, to understand the origin of the 13.6 eV binding energy of the electron in hydrogen. By a rough classical calculation (which works pretty well here), the energy of the electron at some particular distance is just its potential energy (7.13). And the typical distance is the Bohr radius, (7.14). Combining these two equations, we guess that the binding energy should be

$$E_1 \approx \frac{\alpha^2}{\lambda_e} = \alpha^2 m_e = 27.2 \text{ eV}. \qquad (7.15)$$

Oh no, we got the wrong answer! Don't worry. If you stare closely, we actually got twice the right answer. That is not a failure—it's a pretty good result, considering we just used dimensional analysis and some hand-waving. This technique wouldn't be good enough to do precision physics, but it gets us in the right ballpark and provides some intuition for how different quantities depend on each other.

MOLECULES

The ratio in mass between the proton and the electron, $m_p/m_e \approx 1{,}800$, might seem like a hierarchy that begs for explanation. But it's not that big of a number, and some particle has to be the lightest, after all. Likewise, $\alpha = 1/137$ is small, but not crazily small. Maybe that's just the way things are.

Once we have those numbers in hand, however, we can begin to understand the cascade of energy scales that leads us from elementary particles to the everyday world. The binding energy of an electron in hydrogen is of order 10^1 eV, smaller than the electron mass by a factor of α^2. In more massive atoms, the outermost electrons are generally farther away from their nuclei, and their binding energies are smaller yet. Typical numbers are therefore between 1 and 10 eV. For similar reasons, those are also the scales characterizing the energy of binding (or dissociation) of atoms with each other, within molecules. That is to say, those are the energies of chemical reactions. Chemistry arises from atoms sharing (or ceasing to share) electrons with each other, all of which ultimately comes down to the mass of the electron and appropriate powers of the fine-structure constant.

There are also energies associated with molecules themselves. A molecule can vibrate when two atomic nuclei within it move together and apart. Just from the description, you will not be surprised to learn that there are energy levels for those motions that are well approximated by our friend the simple harmonic oscillator. Typical energy scales for molecular vibrations are of order 10^{-2} eV. There are also subtler kinds of molecular motions, in which the overall shape of the molecule gets twisted or distorted. These are somewhat gentler, with energies typically of order 10^{-3} eV.

And so on down the line as we consider complicated molecules, such as we find in organic chemistry. The existence of a large number of ways for a molecule to jiggle means that the lowest-energy examples are likely to be even lower energy than the typical rotations of a simple molecule. An extended, complicated molecule will have vibrations with relatively long wavelengths and long timescales, corresponding to small energies. In this way, chemistry builds up a rich hierarchy of energy scales, all much lower than what we're used to dealing with in particle physics. It will never cease to be amazing how the complexity of our everyday world is constructed from just a few simple elementary ingredients.

EIGHT

SYMMETRY

The very first chapter in *Space, Time, and Motion* was devoted to the idea of "conservation." One of the things we mentioned was Noether's theorem, according to which every continuous symmetry transformation of a system is associated with a conserved quantity. And we said that a symmetry is a transformation you can do to a system that leaves its essential features unchanged.

All true. And symmetries are helpful throughout classical mechanics, especially in relativity, where the fact that the laws of physics don't change their forms when we do a Lorentz transformation is an especially powerful example of symmetry in action. But it's in quantum field theory that symmetries are promoted from "really useful" to "absolutely indispensable." There's a sense in which the forces of nature arise directly from a particular kind of symmetry of the underlying fields, known as "gauge invariance." We'll devote a whole chapter to it, but first we need to wrap our brains around the general idea of symmetry.

And that means we have to think about **group theory**, which is the mathematical framework for classifying and using symmetries. The concepts involved are pretty simple—certainly compared to calculus, or non-Euclidean geometry, or for that matter quantum mechanics—but the language and notation can seem abstract and unfamiliar, so it's worth taking it slowly. To keep our bearings, here is a quick outline of what we'll be seeing in this chapter:

- Symmetries are transformations we can do on things—whether they be geometric shapes or sets of quantum fields—that leave the essence of the things unchanged.
- Simple examples are how we can rotate or flip shapes like triangles and squares. These are "discrete" symmetries, as there is a discrete set of transformations we can do.
- Sets of symmetries are described mathematically as "groups." By thinking of the properties of groups in general, we can gain powerful insights into how symmetries function in physics.
- There are continuous as well as discrete symmetries, like the set of rotations in a certain number of dimensions. Rotations in an n-dimensional space are described by the orthogonal groups $SO(n)$.
- There are also spaces made of n complex dimensions, useful when describing charged particles. Then rotations are described by the unitary groups $SU(n)$. There are a lot of mathy details associated with unitary groups, but the basic idea is just "rotations of vectors made from complex numbers."

The reason why the Standard Model of particle physics is sometimes known as "$SU(3) \times SU(2) \times U(1)$" is because that's the label for the specific set of gauge symmetries that appear (as far as we know) in the real world. Let's set about figuring out what that means.

INVARIANCE

Human beings tend to view the world in human terms. Our ideas about symmetry, for example, are influenced by how we look at other human beings. According to some anthropological evidence, we tend to judge symmetric facial features as generally more attractive. In this case, the relevant symmetry is bilateral: we imagine reflecting a face around a vertical line down the middle, and we judge it to be symmetric if the two sides are nearly mirror images of each other. (A face that was symmetric with respect to a horizontal line drawn in the middle would seem hideous.) Or, more casually, we think of symmetry as an aspect of the overall balance or proportion of a person or object.

But bilateral symmetry is just one example of a much more general concept. In physics or mathematics we often consider the idea of a **transformation** of an object—some process (mathematically, a map or function) that changes the object in some way, either intrinsically or in relation to the rest of the world. Picking up and moving an object is a transformation, but so is distorting its spatial configuration or reflecting it around some plane. We need not physically carry out any particular transformation in order to inquire about what its properties would be were we to do so.

If a transformation leaves some property unchanged, we say that the property is **invariant** under the transformation. And when there is such an invariance, the corresponding transformation is officially a **symmetry**. We will ultimately be interested in symmetries obtained by transforming quantum fields into one another. The strong interactions, for example, are based on a symmetry labeled $SU(3)$ that rotates the three quark colors into each other. But we can start more concretely by thinking about symmetries of geometric shapes.

Consider the following two-dimensional figures: an amorphous closed curve, an equilateral triangle, a square, and a circle.

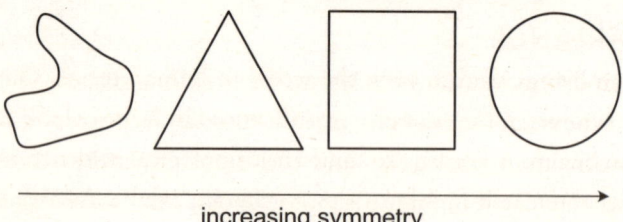

increasing symmetry

The shapes have been placed in order of increasing symmetry. The amorphous curve is the least symmetric, followed by the triangle, then the square, then finally the circle.

For the amorphous curve, this makes perfect sense: not much symmetry there. Likewise, the circle seems pretty clearly the most symmetrical. It's hard to be more symmetrical than a perfect circle.

In the cases of the triangle and the square, the situation is less clear. They both seem pretty symmetrical, maybe equally so. But there is a rigorous sense in which the square has more symmetry than the triangle. To see that, we'll have to be more precise about how symmetries are categorized mathematically.

TRIANGLE SYMMETRIES

Let's consider the triangle in some detail. How much symmetry does it have, and what does it even mean to ask "how much symmetry" there is in a particular situation?

A symmetry is a transformation that leaves an object invariant. So we have to be super careful about what is the "object" we're considering, and what kinds of transformations we're interested in. For the triangle, let's label its corners so we know what we're doing to it: the top corner is A, then moving clockwise we have B in the lower right corner and C in the lower left. "The triangle" is the pure geometric

SYMMETRY

figure itself, not the labels; they just help us keep track of how we are transforming it.

If we rotate the triangle clockwise by 120 degrees (one-third of a full rotation), the corner A moves to where B was, B moves to where C was, and C moves to where A was. The resulting shape is identical to the shape we started with. So this 120-degree clockwise **rotation** is a transformation, labeled R_+, and that transformation is indeed a symmetry. (We're using the + sign to denote "clockwise.") Rotating clockwise by 90 degrees is also a transformation N_+ that we can do on the triangle, but it is not a symmetry, since the triangle obviously looks different after we do it.

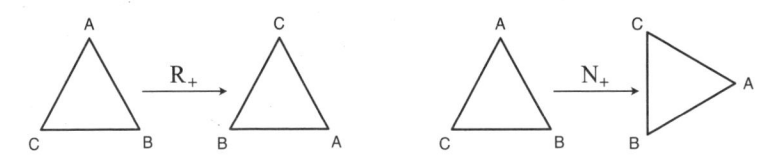

Now we can ask, what are *all* of the symmetries of the triangle? Clearly we can rotate clockwise by 120 degrees. It will come as no surprise that we can also rotate clockwise by 240 degrees (two-thirds of a full rotation). Or, for that matter, by 360 degrees (a full rotation). Or any other multiple of 120 degrees. Furthermore, we could also rotate in the other direction (counterclockwise) by 120 degrees; let's call that transformation R_-. And we can rotate counterclockwise by any multiple of 120 degrees as well. All of these transformations are symmetries, since they leave the original shape invariant.

There are also symmetries of the triangle that aren't rotations at all. Consider reflection around a vertical line, drawn from the top corner and bisecting the triangle. This **flip**, which we can label F_1, has the effect of switching the bottom two corners while leaving the top one unchanged. (It could also be called a "reflection," but the letter R is already reserved for rotations.) Likewise we can consider F_2, a flip around a bisecting line drawn from the lower right corner, and F_3, a flip around a bisecting line drawn from the lower left corner.

By exchanging just two of the corners, any one of the flip transformations changes the **parity** or "handedness" of the triangle. Going clockwise from A, we originally would reach B and then C, and that fact is unchanged by doing rotations. But if we perform any one of the flip operations, going clockwise from A takes us to C and then B. That's how we can be sure that no combination of rotations is equivalent to a flip. (When we're confined to two dimensions anyway.) All the transformations we've considered are portrayed in the figure, starting from the original triangle in the top center.

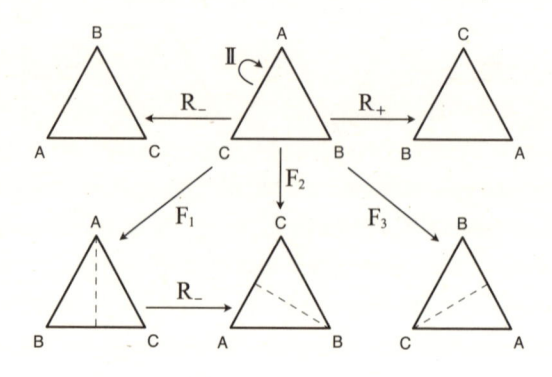

You have undoubtedly noticed that these transformations are not independent of each other: one transformation can be equivalent to a combination of others. For example, doing R_+ twice (that is, rotating clockwise by 240 degrees) is equivalent to doing R_- (counterclockwise by 120 degrees) just once. In either case, we end up with corner A in the bottom left, B on top, and C in the bottom right. The triangle doesn't remember, or care about, which direction it moved in; all that matters is the final configuration of the shape after doing the transformation. Rotating by 360 degrees in either direction is therefore equivalent to doing nothing at all. We have to keep track of every

SYMMETRY

possible symmetry transformation, even one this trivial—"doing nothing at all"—so it is labeled the **identity transformation**, or simply $\mathbb{I}$. And as we can see from the figure, doing an F_1 flip followed by an R_- rotation is equivalent to doing an F_2 flip.

GROUPS OF SYMMETRIES

You can play around with other combinations of our symmetry transformations and see what you get. Even better: convince yourself that we aren't missing any symmetries. Everything you can do to an equilateral triangle that leaves it looking like exactly the same equilateral triangle can be thought of as either a rotation, a flip, or the identity (keeping the original triangle unchanged). This complete set is called a **group** of symmetries. In the case of the equilateral triangle, the symmetry group is known as the "dihedral group of degree 3," which sounds way more intimidating than our collection of rotations and flips actually is. It's written D_3, and it consists of the six elements

$$D_3 = \{\mathbb{I}, R_+, R_-, F_1, F_2, F_3\}. \tag{8.1}$$

Nobody should be upset if you just call it "the triangle group."

The reason this is the dihedral group of "degree 3" derives from the triangle having three corners, which are exchanged with each other under action of the group. The symmetry group of a square would be D_4, for an octagon it would be D_8, and so on. For any regular polygon with n corners, the symmetry group D_n has $2n$ elements: n rotations (including the identity) and n flips. That's why we say the square ($n=4$) has more symmetry than the triangle: there are eight elements in its symmetry group, as opposed to just six for the triangle. You can work them out by playing with rotations and reflections that leave the square invariant. Hint: there are flips around axes that go from one corner to the opposite one, and also flips that go from the middle of one side to the opposite one.

Our triangle is a pretty simple thing. You can probably visualize what's going on in your head when we flip or rotate it. Needless to say, there will be more complicated and subtler symmetries for which visualization will fail us. When first encountering complicated ideas in physics or math, people often ask, "How do you visualize that?" The short answer is, "You don't." The longer answer is, "You can try to visualize simplified lower-dimensional analogues of what we're talking about, but ultimately you are constructing some mathematical structure and trying to figure out what the equations are implying, rather than directly visualizing what you have in mind." (The same might be said for quantum mechanics and field theory more generally.) The good news is that math is pretty reliable, if you use it correctly.

It would be nice to have some notation for combining symmetries together, for example, to indicate that an F_1 flip followed by an R_- rotation is equivalent to doing an F_2 flip. Here's how we're going to write that fact:

$$R_- \cdot F_1 = F_2. \tag{8.2}$$

There are a couple of things going on here. On the left it looks like we are multiplying R_- by F_1, and indeed mathematicians will sometimes talk that way. But it's not exactly the same as the usual idea we have in mind when we're multiplying numbers. By "$R_- \cdot F_1$" we mean "first do the flip F_1, and then the rotation R_-." That's right-to-left, which might look backward, but in our minds we are thinking of these as transformations that act on the triangle. By thinking of our transformations as functions, you can understand (8.2) as expressing $R_-(F_1(\Delta)) = F_2(\Delta)$, where Δ indicates the triangle under consideration.

Most important, these transformations don't **commute**, which is

SYMMETRY

a way of saying that the order in which we write the transformations matters. Looking back at the figure, you can deduce that doing these transformations in the opposite order, R_- and then F_1, would give us a triangle with B at the top, A at the bottom right, and C at the bottom left. That's equivalent to performing F_3 on the original triangle, not F_2. So we would write $F_1 \cdot R_- = F_3$. Combining all of our group elements in various ways, we could work out a "multiplication table" (even though it's not usual multiplication) that would summarize the entire structure of the group.

Here we've explicitly written out the group multiplication table for D_3. The convention is that in the group operation, the element labeling the column goes on the right (and therefore acts first) and the element labeling the row goes on the left. Looking at the F_1 column and R_- row, we have $R_- \cdot F_1 = F_2$ as expected.

D_3	$\mathbb{I}$	R_+	R_-	F_1	F_2	F_3
$\mathbb{I}$	$\mathbb{I}$	R_+	R_-	F_1	F_2	F_3
R_+	R_+	R_-	$\mathbb{I}$	F_3	F_1	F_2
R_-	R_-	$\mathbb{I}$	R_+	F_2	F_3	F_1
F_1	F_1	F_2	F_3	$\mathbb{I}$	R_+	R_-
F_2	F_2	F_3	F_1	R_-	$\mathbb{I}$	R_+
F_3	F_3	F_1	F_2	R_+	R_-	$\mathbb{I}$

Once we become elite mathematicians, we tend to stop thinking of groups as "sets of symmetry transformations on objects" and start thinking of them as "sets of things we can combine together in a specified order to obtain another element of the set." There are some restrictions, or **group axioms**, on when such a set qualifies as a group—mainly that one unique element must qualify as the identity and that every element has an inverse. To say that "there exists an identity element $\mathbb{I}$" means that, for any given group element g, that element is left unchanged by acting with $\mathbb{I}$ on either the right or the left:

$$g \cdot \mathbb{I} = \mathbb{I} \cdot g = g. \tag{8.3}$$

And "every element has an inverse" means that, given g, there exists an element g^{-1} that gives the identity when we multiply it by g on either side:

$$g \cdot g^{-1} = g^{-1} \cdot g = \mathbb{I}. \tag{8.4}$$

In words, in every group there is always a transformation that leaves the object unchanged, and any transformation in the group can be undone by some other transformation.* The inverse of an element might be the element itself; that's true, for example, of each of our flips in the triangle group.

GROUP THEORY

In the eyes of a mathematician, a group is just a set with a particular kind of structure: basically, a "multiplication rule" that has an identity element and an inverse. This idea plays a special role in the study of symmetries, because in the set of symmetry transformations there is indeed always an identity element (don't do anything) and an inverse to every transformation (undo what you originally did).

---

* Strictly speaking, there are two more group axioms. First, closure: combining any two group elements yields another group element. And second, associativity: for any three group elements a, b, and c, we have $(a \cdot b) \cdot c = a \cdot (b \cdot c)$.

SYMMETRY

One of the obstacles to learning group theory for the first time is that the ideas of "multiplication" and "having an inverse" sound familiar to us from how we manipulate ordinary numbers, but there are crucial distinctions that can't be ignored. Consider the set of real numbers, and for our multiplication rule let's take good old-fashioned multiplication: $2 \times 3 = 6$, that kind of thing. Immediately we discover that this is *not* a group, because the requirement that every element have an inverse is not satisfied: 0 has no inverse. There is an identity element, namely 1, but that's not enough to save it. To be a group, you have to satisfy all the group axioms without exception.

When we think about a would-be group, it's important to specify not only the elements of the set but the specific binary "multiplication" operation we have in mind. The real numbers under ordinary multiplication are not a group, but think about the **real numbers under addition**: $2 + 3 = 5$ and so on. That *is* a group, which we label $\mathbb{R}$, and sometimes lazily call "the real numbers" or simply "the reals." In this group, the inverse of x is just $-x$, and the identity element is 0. The "multiplication rule" is what you grew up thinking of as addition. You just have to get used to that.

Now we are off to the races, as far as group theory is concerned. Given the axioms we have specified for defining a group, we can begin to ask all sorts of questions about how groups work and how different groups relate to each other. For example, two groups are said to be **isomorphic** to each other if they are really the "same" group, even if their definitions seem different. An isomorphism between two groups is a one-to-one correspondence between the elements of the two sets, which has the property that the multiplication rule is preserved by this correspondence. That is, multiplying in one group and then mapping to the other gives us the same answer as mapping from one to the other and then multiplying there. If two groups G and H are isomorphic, we write that as $G \cong H$.

Then we have the crucial notion of a **subgroup**: a subset of the set of group elements that forms a group all by itself, with respect to the same multiplication rule. In practice this amounts to checking for closure: in a subgroup, we have to be able to combine any two elements and still get an element of the subgroup, without jumping out into the bigger group. In the triangle group, the set of rotations $\{\mathbb{I}, R_+, R_-\}$ is a subgroup—the combination of any two rotations is still a rotation. But something like $\{\mathbb{I}, R_+\}$ is not a subgroup, since $R_+ \cdot R_+ = R_-$, which isn't in the list of elements $\{\mathbb{I}, R_+\}$.

INTEGERS

The **integers**, labeled $\mathbb{Z}$, are an obvious subgroup of the reals, $\mathbb{R}$. The elements are the infinite set $\{\ldots -1, -2, 0, 1, 2, 3, \ldots\}$, and addition is once again the appropriate binary operation. (So really we should call this group "the integers under addition," but you are expected to fill in the implicit conditions.) It is traditional, for the reals or the integers or any other group where we combine elements via addition, to use the usual $+$ sign to denote the binary operation, rather than the center dot $\cdot$ sign. Note that we started off, in our discussion of the triangle group, thinking of group elements as transformations of an object. Now with the reals or integers, we are thinking of group elements as objects in their own right: good old numbers. That's okay. As far as group theory is concerned, they're all just elements of some set, on which we define some binary operation.

Here's a question to contemplate: given that they both have an infinite number of elements, are $\mathbb{Z}$ and $\mathbb{R}$ isomorphic to each other? (Hint: no, they are not. Not all infinities are created equal.)

When thinking about subgroups, it's important to remember that groups are sets of elements with a certain binary operation. Just because we have a sub-*set*, doesn't mean we have a sub-*group*. We have to check that the subset is closed under the group operation. You might

SYMMETRY

think, for example, that the whole numbers—all integers zero or greater—would be a good candidate for a subgroup of the integers, but it doesn't quite work. While 0 could function as an identity element under addition, there are no inverses, since there are no negative whole numbers. So the whole numbers are not a group at all, much less a subgroup.

We do have finite-sized groups called the **integers modulo n**, or $\mathbb{Z}_n$ for short. These groups have n elements, $\{0, 1, 2, \ldots n-1\}$. The group operation is addition as before, but now "modulo n," or simply "mod n." That means if p and q are in $\mathbb{Z}_6$, and we add $p + q$, if the answer is 6 or greater we simply subtract 6 from the answer. Therefore, addition mod n will always give us a result between 0 and $n-1$. This is subtly different from the ordinary addition operation on the integers as a whole, so $\mathbb{Z}_n$ is not a subgroup of $\mathbb{Z}$. The elements of $\mathbb{Z}_n$ can be thought of as discrete points arranged on a circle: once we count all the way up to $n-1$, we return back to 0 to start again.

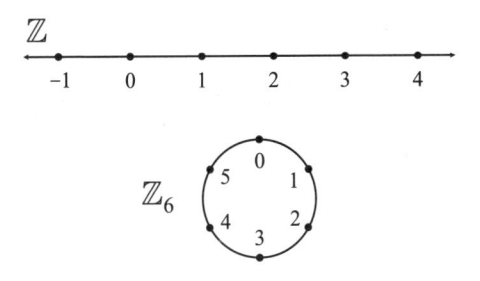

Let's think about $\mathbb{Z}_6$, the integers modulo 6. It has six elements: $\{0, 1, 2, 3, 4, 5\}$. That's the same number of elements as our triangle symmetry group D_3. It's natural to wonder whether $\mathbb{Z}_6$ is isomorphic D_3, but it is not. The multiplication table is different, so they're not the same groups. It's actually pretty easy to see this, once we remember that transformations in the triangle group do not commute with each other. For example, we noted that $R_- \cdot F_1 \neq F_1 \cdot R_-$. But addition, whether in the full integers $\mathbb{Z}$ or a group like $\mathbb{Z}_6$, certainly does commute: $p + q = q + p$ for all p and q. So the group operation cannot work the same way in $\mathbb{Z}_6$ as it does in D_3.

Groups where the operation always commutes, $a \cdot b = b \cdot a$ for all elements a and b, are called **abelian**, after Norwegian mathematician Niels Henrik Abel. Groups where the operation sometimes doesn't commute are, naturally enough, **non-abelian**. $\mathbb{R}$ and $\mathbb{Z}$ and $\mathbb{Z}_n$ are all abelian groups, while D_3 is non-abelian. There are important examples of both kinds of groups in physics, but non-abelian groups are obviously more complicated and therefore more fun.

It's important to internalize the ideas that (1) a group is defined by both its set of elements and the binary operation defined on that set, and (2) two groups that look different are really "the same," as far as their groupness is concerned, if there is an isomorphism between them. Consider $\mathbb{Z}_2$, the integers modulo 2. As we've defined it, the elements are $\{0, 1\}$, and the operation is addition mod 2, with 0 as the identity element. But we can also consider something called the "multiplicative group of order 2," with elements $\{+1, -1\}$, and the operation given by multiplication, so that the identity element is $+1$. This group is isomorphic to $\mathbb{Z}_2$, and most physicists will just call it "$\mathbb{Z}_2$" (to the consternation of exacting mathematicians). Keeping this isomorphism in mind, it makes sense to say that "$1 + 1 = 0$" and "$-1 \times (-1) = +1$" are both true, and indeed they express the same idea; the first is just phrased in the language of addition mod 2, while the second is phrased in the language of multiplication.

$\mathbb{Z}_2$ appears all the time in physical situations, since we often have transformations that undo themselves, such as taking the mirror image of something or exchanging particles with antiparticles. For example, the triangle group has subgroups consisting of the identity and any one of the flip transformations, such as $\{\mathbb{I}, F_1\}$. All of these subgroups are isomorphic to $\mathbb{Z}_2$, and most physicists are happy to refer to them as $\mathbb{Z}_2$, or as "a $\mathbb{Z}_2$ subgroup." ($\mathbb{Z}_2$ is not a subgroup of the integers, but it is a subgroup of D_3, because in the latter case we have elements that give the identity when squared.)

CIRCLE SYMMETRIES

We did a good job with the equilateral triangle, completely classifying all of its symmetry transformations, boiling them down to the six-element group D_3. It wouldn't be that hard to do a similar analysis of the square and its eight-element symmetry group D_4. What about the circle?

We know a circle when we see it, but now we're going to need to be more precise. The easiest way to define a circle is to start with a flat, two-dimensional plane, construct perpendicular x and y coordinates on it, and consider the set of all points a distance 1 from the origin of those coordinates. The resulting set of points forms a circle of radius 1. It doesn't really matter that the central point is the origin, or that the radius is precisely 1, but those choices give us a pleasantly definite starting point.

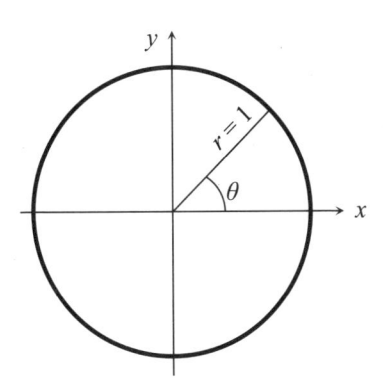

Sometimes a circle is called S^1, as it is a special case of the more general idea of an n-dimensional sphere S^n, defined as "the set of all points a fixed distance from the origin in $(n+1)$-dimensional Euclidean space." Our usual notion of a sphere, like the surface of a basketball, is S^2.

Note that it is only the two-dimensional surface of the sphere we have in mind; the three-dimensional interior is called a *ball*. There is even a zero-sphere, S^0, which is the set of points a fixed distance from the origin on a one-dimensional line. That is, just the two points at r and $-r$ for radius r. This is the kind of generalization beyond intuition that mathematicians love—a zero-dimensional sphere is just a set of two points.

The symmetries of the circle are not that different from those of the triangle: there are rotations and flips. The difference is that there are infinitely many of them. We can rotate the circle by any angle θ between 0 and 2π, and likewise we can reflect the circle about any straight line drawn through the origin. (We're expressing angles in terms of radians, where 2π radians equals 360 degrees.) Denote a rotation by θ as $R(\theta)$, and a reflection about an axis at angle ϕ by F_ϕ. Then the group of symmetries of the circle has elements

$$O(2) = \{R(\theta), F_\phi\}. \tag{8.5}$$

The notation O(2) stands for the **orthogonal group** in two dimensions, which comes from thinking about the space in which the circle is embedded rather than the circle itself. The two dimensions in question are the x and y axes of the two-dimensional plane. Those two axes are orthogonal (perpendicular) to each other, and that property of orthogonality is preserved whenever we either rotate or flip the plane while keeping its origin fixed, whether there's a circle there or not. "The set of transformations of a plane that keep axes perpendicular and the origin fixed" is exactly the same as "the set of symmetries of the circle," and analogous statements will continue to be true in more than two dimensions. Such groups are therefore called "orthogonal groups" rather than "sphere groups," although the latter would also have worked.

The fact that elements of O(2) can be divided into rotations and flips is important, but also somewhat clumsy. Often it is convenient to concentrate on just the subgroup representing rotations. This is known as the **special orthogonal group** in two dimensions, SO(2), with elements

$$SO(2) = \{R(\theta)\}. \tag{8.6}$$

If you think about it, this group is itself a circle—the rotations for any angle θ between 0 and 2π describe a one-dimensional set of objects that topologically wrap back to their starting point, $R(2\pi) = R(0)$. But it's not going to generally be true that the set of rotations in n dimensions will form an $(n-1)$-dimensional sphere—that's just a coincidence we have when $n=2$.

That brings up an important point. Back when we discussed general relativity in *Space, Time, and Motion*, we introduced the notion of a **manifold**, which is a kind of space that looks like ordinary smooth n-dimensional space when we zoom in sufficiently closely, although it might have complicated geometry or topology when we zoom back out. The groups O(2) and SO(2), which contain an infinite number of elements that smoothly vary from one to another, are themselves manifolds as well as groups. Such groups are named **Lie groups**, after Norwegian mathematician Sophus Lie. (What can I say, nineteenth-century Norway was something of a group-theory hotbed.) They can be contrasted with **discrete groups** like D_3 or $\mathbb{Z}$, where the elements look like disconnected individual points. Both kinds are important in physics, but it's with Lie groups that things really start getting profound.

ORTHOGONAL GROUPS AND VECTOR SPACES

As we mentioned, the symmetry group of the circle O(2) is called an "orthogonal group" because it can equally well be thought of as "the symmetries of a two-dimensional plane, keeping the origin fixed and the axes orthogonal to each other." A two-dimensional plane is an example of a vector space. In actual physics applications, thinking in terms of symmetries of vector spaces will be much more relevant than the intuitive idea of the symmetries of a circle. We will have sets of quantum fields that define vector spaces of various dimensionalities; for example, quark fields live in a three-dimensional "color" vector space with axes labeled by {red, green, blue}. And many of those field values are going to be described by complex numbers rather than real ones. So let's think about vector spaces in a more systematic way.

We've seen **vector spaces** before: sets of elements, called vectors, that can be added together and scaled by numbers. Every vector space has an origin, the zero vector. That acts as an identity element when we think about adding vectors together. So every vector space is itself a group—but our interest here is groups that transform the vector spaces, not thinking of vectors themselves as group elements.

Back in *Space, Time, and Motion*, we generally dealt with real vector spaces (in the sense of real versus complex numbers, not in the sense of physically existing). If we have a three-dimensional vector space with basis vectors $\{\vec{e}_1, \vec{e}_2, \vec{e}_3\}$, a general vector can be written $\vec{v} = v_1\vec{e}_1 + v_2\vec{e}_2 + v_3\vec{e}_3$, where v_1, v_2, and v_3 are all real numbers, called the components of the vector in the appropriate direction. In other numbers of dimensions, we simply have more (or fewer) basis vectors and components: one each for each dimension of the vector space. When we say that vectors are things that can be "added together and scaled by numbers," in a real vector space the numbers we can scale by are real numbers. If $\vec{v}$ and $\vec{w}$ are any two vectors and a and b are real numbers, the combination $a\vec{v} + b\vec{w}$ is also a vector.

Plenty of spaces are not vector spaces. A sphere is not a vector space, nor is the kind of curved manifold we have in general relativity. In both cases there is no way to add two points together, nor is there a preferred point called "the origin." Vector spaces are smooth sets of

SYMMETRY

points that extend infinitely far from an origin in a flat geometry, generalizations of lines and planes.

When we have two vectors $\vec{v}$ and $\vec{w}$, another kind of operation (defined on some vector spaces, not necessarily all) is the **inner product** (or "dot product"), given by

$$\vec{v} \cdot \vec{w} = v_1 w_1 + v_2 w_2 + v_3 w_3. \tag{8.7}$$

If the angle between the two vectors is θ, this formula turns out to be equivalent to

$$\vec{v} \cdot \vec{w} = |v||w|\cos\theta, \tag{8.8}$$

where $|v|$ and $|w|$ are the lengths of the respective vectors. When two vectors are orthogonal, the angle between them is $90° = \pi/2$ radians, and the cosine is 0. So the inner product between two orthogonal vectors always vanishes.

One way of thinking of the orthogonal groups $O(n)$ is that they are transformations in an n-dimensional vector space that preserve the inner product between any two vectors. That implies "keeping the origin fixed," because if we move the origin, the length of the two vectors will change. So orthogonal transformations can be thought of as rigid rotations and reflections of n-dimensional vectors.

The group $O(3)$, for example, is the set of transformations we can do in three dimensions that preserves the inner product (8.6)—in other words, that keeps axes orthogonal and the origin fixed. Equivalently, it is the symmetry group of the two-sphere S^2. This group also consists of rotations—independently around any of the three axes in three-dimensional space—as well as reflections. There is a pure-rotation subgroup, the special orthogonal group $SO(3)$, and once again a set of $\mathbb{Z}_2$ groups that invert the orientation of the space.

SO(2)   SO(3)

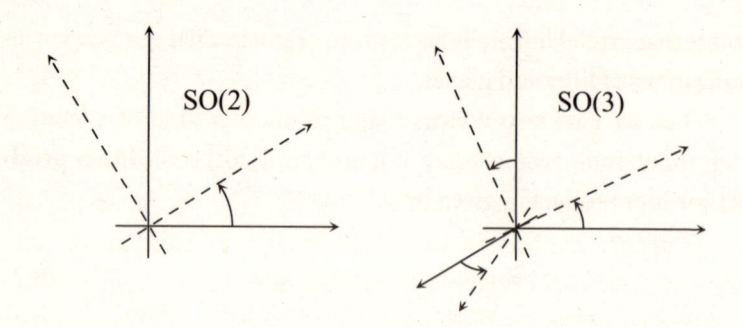

In general, both O(n) and SO(n) will be Lie groups with dimension ½ $n(n-1)$, so SO(3) is three-dimensional.*

SO(2), rotations in two dimensions, is an abelian group. It doesn't matter whether I first rotate by an angle θ_1 followed by an angle θ_2, or the other way around; the total result is a rotation by $\theta_1 + \theta_2$. But all of the other SO(n) for $n>2$ are non-abelian. In three dimensions it *does* matter whether I first rotate around the x-axis and then around the y-axis, or vice versa. This is why non-abelian groups are so important in physics: we often are doing rotation-like things between more than just two axes.

Bonus digression: we've been thinking of elements of SO(n) in terms of what they do, namely rotate things in n dimensions. Often physicists like to make things a bit more concrete and represent such rotations as $n \times n$ matrices that would implement such a rotation by acting on vectors. For example, an element of SO(2) that rotates two-dimensional vectors by an angle θ can be written

---

* Pick an axis to move, and another axis to rotate it toward. That's n axes to start with, and $n-1$ others to rotate it toward. So it looks like $n(n-1)$ ways to rotate, but picking the same axes in the opposite order doesn't count as a separate kind of rotation, so we divide by two. You are welcome to rotate around any line at all through the origin, not just one of the coordinate axes, but an arbitrary rotation can be constructed as a combination of appropriate rotations around the axes, so those are sufficient to define the dimensionality of the group.

SYMMETRY

$$R(\theta) = \begin{pmatrix} \cos\theta & -\sin\theta \\ \sin\theta & \cos\theta \end{pmatrix}. \quad (8.9)$$

In this way of thinking, $SO(n)$ is the set of all $n \times n$ matrices M that satisfy the "orthogonality condition," $M^T M = \mathbb{I}$, where the superscript T denotes the transposition of the matrix (exchange rows with columns). Likewise, elements of the unitary groups we consider below will be matrices satisfying the "unitarity condition," which is the same as the orthogonality condition except we also take the complex conjugate of the first matrix. Happily, we won't ever worry about the explicit matrix forms of any of our transformations. For us, elements of $SO(n)$ are simply "rotations in n dimensions."

COMPLEX NUMBERS REVISITED

Complex numbers are everywhere in physics, and quantum field theory is no exception. The next step in our group-theory journey is therefore to upgrade from the orthogonal groups $SO(n)$, representing rotations in n dimensions, to the unitary groups $SU(n)$, representing rotations in n *complex* dimensions. Which means we need to know what a "complex dimension" is supposed to be. Which gives us the opportunity to take a breath and remember some things about complex numbers.

A complex number is a combination of a real number and an imaginary number, as we talked about in Chapters 1 and 2. If we define the imaginary unit as $i = \sqrt{-1}$, we have an entire set of imaginary numbers that are just regular real numbers times i. Then a complex number is the sum of a real number and an imaginary one:

$$z = a + ib, \quad (8.10)$$

where a and b are themselves real numbers, called the real and imaginary parts of z. When we add complex numbers, the real and imaginary parts are added separately: $(a+ib) + (c+id) = (a+c) + i(b+d)$.

Complex numbers are numbers: you can add them, multiply them, divide by them, and so on. You can think of them and study their properties in their own right, without decomposing them into a sum of real and imaginary numbers. The main difference is that there is a new operation, **complex conjugation**, denoted z^*, which is the inverse of itself: $(z^*)^* = z$. In terms of the real and imaginary parts, complex conjugation puts a minus sign in front of the imaginary part, $z^* = a - ib$. The real numbers remain fixed, while $i^* = -i$.

A nice way to visualize complex numbers is in terms of the **complex plane**, as we briefly saw in Chapter 1. That's just an ordinary two-dimensional plane in which the horizontal axis represents the real part of the complex number, and the vertical axis represents the imaginary part. In this picture, complex conjugation is just a reflection around the real axis, $b \to -b$. The modulus $|z|$, which is just the distance between 0 and z in the complex plane, is a real number satisfying

$$|z|^2 = z^* z = a^2 + b^2. \tag{8.11}$$

Instead of specifying a complex number in terms of its real and imaginary parts, we can alternatively use a "polar representation," in terms of the modulus $|z|$ and the **phase** θ of the number, as

SYMMETRY

$$z = |z|e^{i\theta}. \tag{8.12}$$

The phase is just the angle between the horizontal (real) axis and the line stretching from the origin to z in the complex plane. We can figure it out by remembering Euler's formula,

$$e^{i\theta} = \cos\theta + i\sin\theta. \tag{8.13}$$

If we look at the figure of the complex plane and remember how sines and cosines are used to relate the length of a segment to its horizontal and vertical components, we see that (8.13) is just what we need to make the polar representation (8.12) come out right.

Finally we can get to complex dimensions. A **complex vector space** is exactly the same thing as a real vector space, except that the components are now complex numbers, and likewise the vectors can be scaled by complex numbers. If $\vec{\phi}$ is a vector in a three-dimensional complex vector space, it can be written in terms of components and unit vectors as

$$\vec{\phi} = \phi_1\vec{e}_1 + \phi_2\vec{e}_2 + \phi_3\vec{e}_3, \tag{8.14}$$

where ϕ_1, ϕ_2, and ϕ_3 are complex numbers. (These three dimensions describe some abstract complex vector space, not the real three-dimensional space we live in.) Likewise, for any complex numbers α and β and complex vectors $\vec{\phi}$ and $\vec{\omega}$, the combination $\alpha\vec{\phi} + \beta\vec{\omega}$ is another complex vector. Just like in real vector spaces, we can define a complex inner product, but there is one twist: rather than just multiplying the components in each direction and then adding the results together, we first take the complex conjugate of the components of the first vector. Thus, the inner product of two complex vectors $\vec{\phi}$ and $\vec{\omega}$ can be written

$$\langle \phi, \omega \rangle = \phi_1^* \omega_1 + \phi_2^* \omega_2 + \phi_3^* \omega_3. \tag{8.15}$$

Different references might use other notation than the brackets on the left-hand side. What matters is that we have to keep straight which vector is getting complex-conjugated; the order is important.

Of course, we have already met one important complex vector space: Hilbert space. Wave functions are complex vectors. We generally use special notation for them, like the "ket" notation $|\Psi\rangle$ introduced in Chapter 2, in part to keep reminding us of the particular features of complex vectors.

But the wave function isn't our primary concern right now. The symmetries we care about are going to be symmetries among the fields from which we construct our quantum wave function. We've been talking primarily about a real scalar field, $\phi(x, t)$. But many of the fields that we know about in particle physics are complex rather than real. Now we're in a position to talk about the symmetries of those fields.

UNITARY GROUPS

The set of real numbers is a one-dimensional vector space. Each number is a vector; it can be added to other real numbers or multiplied by some real number. But there aren't "rotations" of just a single real dimension. It's just a line, so what is there to rotate? There is a flip: you can send any real number x to $-x$. That's a single discrete transformation that is its own inverse. So there is an orthogonal group in one dimension, but it's pretty simple: $O(1) \cong \mathbb{Z}_2$.

A single complex dimension is another story. Complex numbers live on a plane, not on a line. And on a plane we can do rotations. The **unitary group** $U(n)$ is the set of transformations of an n-dimensional complex vector space that rotates the vector space around the origin. More formally, it is the set of transformations that leave the inner product (8.15) invariant. Basically those are rigid rotations, just in a

SYMMETRY

complex vector space. When we rotate axes all at the same time, the inner product remains unchanged.*

Unlike O(1), which is just a reflection, U(1) is therefore a set of rotations in the complex plane. Indeed, as groups, U(1) and SO(2), the group of rotations in the real two-dimensional plane, are isomorphic: U(1) $\cong$ SO(2). But they appear different in practice, since U(1) acts on complex numbers and SO(2) acts on two-dimensional vectors. They are both "rotations in a two-dimensional plane," but in the U(1) case we are thinking of two real dimensions as a single complex dimension. U(1) is a one-dimensional abelian Lie group, which we know must be true because SO(2) is.

It's not hard to be explicit about what U(1) does to complex numbers: looking back at our image of the complex plane, an element of U(1) just rotates the phase of the complex number by some angle ω. That's accomplished by multiplying by $e^{i\omega}$. If the number we start with is $z = |z|e^{i\theta}$, the transformed number is $e^{i\omega}z = |z|e^{i(\theta+\omega)}$.

Then we have the **special unitary groups** SU(n), which are just rotations in n complex dimensions. They are (n^2-1)-dimensional nonabelian Lie groups. In matrix form, a general element of SU(2), the simplest special unitary group, can be written

$$M = \begin{pmatrix} a & b \\ -b^* & a^* \end{pmatrix}, \qquad (8.16)$$

where a and b are complex numbers satisfying $|a|^2 + |b|^2 = 1$. SU(2) is supposed to be three-dimensional, and indeed it takes three real numbers to specify such a matrix: the complex numbers a and b are

---

* Technical point: you might think that complex conjugation $z \to z^*$ is analogous to a flip of the complex dimensions and therefore should count as a unitary transformation. It does not; to see this, think about the inner product (8.15). That's a complex number, and in general it will also get conjugated when we conjugate both vectors. But a unitary transformation is defined to leave the inner product completely unchanged, so complex conjugation doesn't count.

equivalent to two real numbers each, and there is one constraint on the sum of their squares, so there are three independent parameters in total. This matrix can be used to rotate vectors in a two-complex-dimensional vector space.

You might think, following what we did with orthogonal groups, that we should first discuss $U(n)$, and then specialize to the pure-rotation case of $SU(n)$, but that's not really necessary. Anytime we want to do a $U(n)$ transformation, that can be implemented as an $SU(n)$ transformation along with a separate $U(1)$ transformation. In practice, physicists will generally analyze $SU(n)$ symmetries and $U(1)$ symmetries separately.

That's it! It was a long journey, but now you know the basics of the symmetry groups that will show up in particle physics. Our favorites will be $SO(n)$, $SU(n)$, and $U(1) \cong SO(2)$. It's a lot of mathematical care just to get comfortable with the idea of "rotations in n (real or complex) dimensions." But it's worth the effort. You can get pretty far as a professional physicist with just those groups.

NINE

GAUGE THEORY

Thus far in the book, we've built up an impressive repertoire of technical know-how about quantum mechanics and quantum field theory. We've learned about wave functions, and why fields lead to particles upon quantization, and how those particles interact with each other, and how to talk about symmetries. In this chapter and the next two, our knowledge-building reaches a crescendo as we grapple with gauge theories and the fermion/boson distinction, two pillars of modern fundamental physics. It will lead to a much-deserved payoff in the final chapter, where we assemble the ingredients of the Standard Model of particle physics and use some simple ideas (conservation laws, heavy particles decaying into light ones) to explain fundamental features of the world we see around us.

This chapter and the next are about **gauge theories**, which are a particular kind of field theory with a particular kind of symmetry—a transformation that can be done independently at every point in spacetime. That simple idea is going to have enormous consequences. We're going to need to introduce extra fields to make this expansive symmetry workable, and those fields are going to end up being the force-carrying fields of particle physics—photons, gluons, W and Z bosons, and gravitons. Gauge symmetry is a powerful principle that undergirds the forces of nature.

When we first talked about how interactions are described by Feynman diagram vertices that can be traced to terms appearing in an interaction Lagrangian, the process may have seemed a little loosey-goosey. Who decides what fields you have, or how they interact? With gauge theories there is much less arbitrariness. Once you have one field and you want it to be invariant under a gauge symmetry, that demands the existence of certain other fields, and how those fields interact is going to be largely determined by the symmetry. It's a reassuringly solid foundation on which to build a physical theory.

There's going to be more mathematical notation in these two chapters than in most of the rest of the book. There is a reason why: you now know enough of the underlying principles to actually see why gauge invariance implies such crucial properties as the masslessness of the photon, the existence of antiparticles, and conservation of charge. It would be a shame to come this far and content ourselves with some informal gestures. It is more fun to put on our thinking caps and come to an honest understanding of some of nature's deepest secrets.

QUARKS AND COLOR

To quickly illustrate what we mean, let's think about a real-world example. **Quarks**, as we discussed in Chapter 7, are the particles living inside protons, neutrons, and other strongly interacting particles. The force that holds them together is described by quantum chromodynamics (QCD), where the role of "charge" is played by a new quantity called "color."

It is often said that quarks come in one of three colors: red, green, and blue. That's not quite accurate. What's happening is that each quark field q is actually, at each point in spacetime, described by a

GAUGE THEORY

vector living in a three-dimensional complex vector space—**color space**—and it's the three axes in that space that are labeled red (R), green (G), and blue (B). Color space is an **internal vector space**—it has nothing to do with velocity or momentum or any other familiar vectors in ordinary space or spacetime. It is a space that simply exists at every point, and each quark field (the up quark, down quark, and so on) defines a particular vector within that space. Put another way, there are three basic colors of quark fields (R, G, B), and something like "the up-quark field at (x, t)" will be a vector combination of all three, with different components in each direction. Color vectors have a direction in color space, not in real space.

You will not be surprised to learn that there is an SU(3) symmetry acting on this three-dimensional complex space. We can "rotate" the quark field with an SU(3) transformation, which changes the relative amounts of R, G, and B components of the field while leaving its length unchanged. "Rotate" is in scare quotes because it's not a literal rotation in space, just a transformation between three directions in this internal color space.

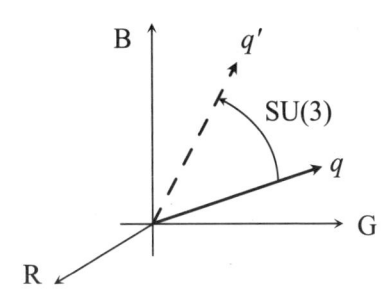

But nature doesn't care about how we express the quark vector in terms of components, or how the axes are oriented—that's just a choice we make for convenience, not a fundamental feature of the universe. That's why there is a symmetry. How we orient the field vector with respect to the axes is physically irrelevant, and symmetries often express the fact that there are multiple equivalent ways of representing the same physical situation. For a single quark field, all that matters is the length of the vector; when we start to compare different fields, or a single field at different points in spacetime, the relative angle between the color vectors will start to matter. Those properties—length of individual vectors, angles between vectors—are precisely what an SU(3) transformation leaves invariant.

GAUGE TRANSFORMATIONS

But that's not the exciting part quite yet. What's important about the SU(3) color symmetry is that it's not a **global symmetry** but rather a **gauge symmetry**, also known as a **local symmetry**. Here is where the magic happens.

Global symmetry transformations are ones that transform the field in precisely the same way at every point in spacetime. They are sometimes called "rigid" symmetries for this reason. Let's denote a particular color rotation by a specific SU(3) group element G. A transformation of the quark fields can be written

$$G: q(x, t) \to q'(x, t) = G \cdot q(x, t). \tag{9.1}$$

The notation on the right-hand side just means "act the transformation G on the field q."

In contrast, a gauge symmetry involves a transformation that happens independently at every point. Rather than a fixed group element G, we consider a spacetime-dependent transformation $G(x, t)$. The rule still looks like (9.1), but the transformation from q to q' can be different at each point.

$$G(x, t): q(x, t) \to q'(x, t) = G(x, t) \cdot q(x, t). \tag{9.2}$$

This is a **gauge transformation**—a symmetry transformation that depends on where you are in space and time.

GAUGE THEORY

This might not seem like such a big deal. Given the basic idea that "the direction of the RGB axes is physically irrelevant," it's natural that it should be separately irrelevant at every point in spacetime.

And it is. But a new issue arises. While the direction in color space in which the field is pointing at any one point is a matter of convention, the *relative* direction in which the field points at two different locations in space can certainly matter. We want to be able to tell whether the quark field at two different points is doing the same thing or doing something different. (For example, we want to know that all of the quark excitations inside a proton have a total color of zero.)

But how are we going to compare what the field is doing at two different points in space if we are free to rotate our axes separately at each point? Given any two locations, we can always do a gauge transformation as in (9.2) to make the quark field point along "blue" at both places, or for that matter make the field red at one point and green at the other. We need some way to really compare the fields at different points—a way that isn't affected by whatever gauge transformations we might choose to do.

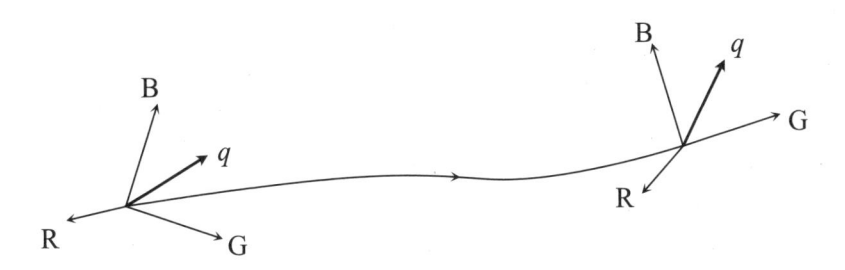

CONNECTIONS

We faced a somewhat similar issue in the curved spacetime of general relativity, but there we were thinking of good old vectors that point in a direction in spacetime rather than the internal vectors of color space. In flat spacetime you could have a vector at one location, and another vector in another location, and simply compare them to each other, at least as long as you sensibly chose to use an orthogonal (Cartesian) coordinate system everywhere. But once spacetime is curved, there is no such thing as an orthogonal coordinate system everywhere. We need to think harder about how to transport a vector from one point to another, along some given path, so that we could bring two vectors to a common point and compare them.

Dedicated readers of *Space, Time, and Motion* might remember that we can indeed do that, and the answer is going to involve another kind of field. There is an **equation of parallel transport** that allows us to move vectors (or other tensors) along any path we like, keeping them as constant as we can along the way. But you need some mathematical information to tell you how the components of the vector would properly change during the process. That information comes in the form of a **connection**, which is itself a kind of field. In general relativity, as it happens, the connection can be derived from the metric tensor field, which is why we didn't have to specify it separately. It is possible to think of general relativity as a gauge theory, as first discussed by Ryoyu Utiyama in 1956, with the gauge group being the Lorentz group $SO(3,1)$. ("Rotations" in a spacetime with three spacelike dimensions and one timelike one.)

Here is a possibly helpful analogy: Imagine you are building a robot whose job will be to work in a restaurant, carrying a tray laden with glasses full of drinks. You want to ensure that the glasses are always oriented exactly vertically so that the robot never spills a drop. But the restaurant is outdoors, and on hilly terrain, so the tray needs to be constantly adjusted so that it remains horizontal. And (don't look at me, I just make up the analogies) the robot won't have any sensors or other devices that would allow it to judge the height and angle of the ground below it. Instead, you are going to give it that information ahead of time: the precise topography of the region over which the robot might travel, so that it will know how to continually adjust

GAUGE THEORY

its tray to keep the drinks from spilling. That information you are feeding the robot is essentially a connection: it's a field (information about something happening at every point in space) that allows the robot to implement the idea of "keeping the tray precisely horizontal, and therefore the drink glasses precisely vertical" as it moves around. The connection field is precisely the data that "connects" one point to what's going on at other points.

The same kind of story is going to hold true in our gauge theories with internal vector spaces. In order to compare quark field values at different locations, we need to introduce a **gauge field**, also called the **vector potential**, which will allow us to move fields around in a gauge-invariant way. "Vector potential" comes from the fact that it's a vector field, generally written $A_\mu(x,t)$. So instead of just a lone quark field, in this gauge theory we need two fields:

$$q(x,t), \ A_\mu(x,t).$$

The push and pull between these two different kinds of fields—matter fields like quarks and leptons, and gauge fields like gluons and photons—creates nucleons, atoms, molecules... all the way up to you and me.

Mathematically speaking, the gauge field is a connection, much like what we used to do parallel transport in general relativity. It provides the information you need to say, for example, "the quark field is constant along this path," even while gauge transformations allow you to change the orientation of your red/green/blue axes however you like from point to point.

There are crucial and fun complications when we have non-abelian gauge symmetries like SU(3), rather than an abelian one like U(1). Then the gauge field A_μ isn't just a vector—it is actually a matrix-valued vector, which has extra indices that allow it to perform rotations in color space or whatever other internal vector space we're considering. For simplicity's sake we are not going to write those out, and instead just use A_μ for whatever gauge field we're talking about at the moment. The fundamental ideas remain the same.

If all this seems a bit abstruse, just remember the bottom line: to have gauge invariance, we need a new field, the gauge field. If you don't want to sweat parallel transport and connections and internal vector spaces, just hold on to the necessity of making sure everything is truly invariant under the gauge symmetry.

Once we have recognized the need for a gauge or connection field, there's every reason to think about that new field in its own right. What are its dynamics? What kinds of particles does it lead to when we quantize? How do those particles interact with particles from other fields?

Every gauge symmetry comes with its own connection field, and those fields and their interactions are what give rise to the known forces of nature. The particles associated with those fields are called **gauge bosons**. For color SU(3), the gauge bosons are gluons. For electromagnetism, they are photons. For the weak interactions they're the W and Z bosons—but as we'll see, there are new complications there because of the Higgs boson and its trickery. Nobody ever said particle physics would be tidy.

Symmetries aren't just a convenient simplification or an attractive aesthetic feature of quantum fields. Implementing them consistently leads directly to forces between particles of matter. That idea is at the heart of the Standard Model of particle physics.

GAUGE INVARIANCE

It will be helpful to take a step back from the complications of QCD and look at a simpler case: QED, quantum electrodynamics, the theory of electrons and positrons and photons. As in QCD, there is a gauge field A_μ that helps us to compare fields at different points. That

GAUGE THEORY

fact has crucial implications for what properties this field has and how it interacts.

So let's take a deep breath and think about how this field is supposed to behave when we do a gauge transformation. Short version: the job of A_μ is to ensure that everything remains gauge-invariant, and that requirement will determine how A_μ itself transforms. And those transformation properties will imply that the physically relevant ("gauge-invariant") features are actually derivatives of A_μ, and those derivatives turn out to be the good old electric and magnetic fields (or their generalization to other forces).

Take an electrically charged field like the electron. We'll denote it by $\psi_e(x, t)$, since the letter e all by itself is already taken by Euler's constant, which we'll be using a lot. (But ψ_e here is a field, not a wave function.) It is a complex-valued field, not a real one. Because electrons are spin-½ particles, it is not a scalar field but something called a **spinor**—that leads to all sort of fiddly complications, which we will cheerfully ignore.

What matters to us is that ψ_e is a complex field. That allows it to have a U(1) gauge symmetry, and indeed it does.

$$\text{U(1):} \ \psi_e(x, t) \to \psi'_e(x, t) = e^{i\theta(x,t)} \psi_e(x, t). \tag{9.3}$$

All we do is multiply the complex field by a spacetime-dependent phase factor, rotating the field value in the complex plane by an angle $\theta(x, t)$. We want to insist that the theory as a whole be invariant under that kind of symmetry transformation. We are therefore going to need the introduction of a gauge field $A_\mu(x,t)$. This will act as a "potential" for the electric and magnetic fields, and quantizing it will lead to photons, so we can also call it the **photon field**.

The gauge field A_μ is a real-valued field, not a complex one. Nevertheless, in order for it to do its job (telling us how to parallel-transport the electron field from place to place), it will also have to change when we do a gauge transformation. If it weren't for gauge symmetry, the simple idea of "the electron field is constant everywhere" would be implemented by setting its partial derivatives with respect to the coordinates equal to 0 (using $\partial/\partial x^\mu = \partial_\mu$ notation):

$$\partial_\mu \psi_e = 0. \tag{9.4}$$

Straightforward enough, but there's an immediate problem when we do a gauge transformation. If we let $\psi_e \to e^{i\theta}\psi_e$ and then take the derivative, we get (keeping the derivative of ψ_e itself 0)*

$$\partial_\mu \psi_e \to \partial_\mu(e^{i\theta}\psi_e) = (\partial_\mu e^{i\theta})\psi_e + e^{i\theta}(\partial_\mu \psi_e) = i(\partial_\mu \theta)e^{i\theta}\psi_e. \tag{9.5}$$

We see that the gauge-transformed derivative of the electron field does not stay 0 but is proportional to the derivative of θ. So it would vanish if θ itself were constant through spacetime ($\partial_\mu \theta = 0$), but in that case we would just have a rigid global transformation rather than a spacetime-dependent gauge transformation. We only have a true symmetry if the quantity is completely unchanged under the transformation. How can we fix things up so that we still have a symmetry when θ changes from place to place?

In a gauge theory, the simple equation (9.4) is not what we mean by "keep the field constant." As we said, we need to use the gauge field to make sense of that statement, since we need to compare values of ψ_e at different locations. The way to do that is to replace (9.4) by

$$\partial_\mu \psi_e - iA_\mu \psi_e = 0. \tag{9.6}$$

This isn't an equation that's supposed to always be true; it's just the gauge-theory generalization of the condition $\partial_\mu \psi_e = 0$. Rather than

---

* Remember the rule for the derivative of a product: $\partial_\mu(fg) = (\partial_\mu f)g + f(\partial_\mu g)$.

GAUGE THEORY

setting the partial derivatives of the field to 0, they need to be compensated by the field times the vector potential. The expression on the left-hand side is going to play a crucially important role in what's to come. It's sometimes called the **gauge-covariant derivative** of ψ_e, but that ponderous nomenclature is less important than the underlying idea of correcting a derivative by the connection. (Think of keeping the field constant as "holding the tray horizontally," and the second term as "how to compensate for the fact that the ground is tilted.")

The benefit of (9.6) over (9.4) is that it is left unchanged by a gauge transformation. In order to make that happen, we need to have a specific kind of transformation rule for A_μ itself:

$$\text{U(1):} \quad A_\mu(x,t) \rightarrow A'_\mu(x,t) = A_\mu(x,t) + \partial_\mu \theta. \tag{9.7}$$

So while the electron field is multiplied by $e^{i\theta}$, the photon field is shifted by $\partial_\mu \theta$, the partial derivative of θ with respect to the spacetime coordinates. This is going to have crucial consequences down the line. And it kind of makes sense. The job of A_μ is to compensate for a spacetime-dependent gauge transformation. But if θ is just a constant, that's a global rotation that doesn't change from place to place. In that case the derivative vanishes and A_μ remains unchanged. So whatever change it undergoes is going to depend on $\partial_\mu \theta$, and the right change turns out to simply be additive.

Now look back at (9.6). When we do a U(1) gauge transformation, the electron field is multiplied by $e^{i\theta}$ as in (9.3), so its derivative changes as in (9.5). But the photon field also transforms, shifting by $\partial_\mu \theta$ according to (9.7). The result is that (9.6) is just multiplied by an overall phase $e^{i\theta}$ after a gauge transformation. So if this quantity is 0 to start, it remains 0.

When James Clerk Maxwell wrote down the equations for classical electromagnetism in the 1800s (standing on the shoulders of giants, to be sure), he did so in terms of the electric field $\vec{E}$ and the magnetic field $\vec{B}$, both of which are ordinary three-dimensional vectors. How do these relate to the gauge field that we are talking about? The answer is that they are derivatives of it, but in a very specific way. If you will forgive a brief relapse into the tensor language we talked about in *Space, Time, and Motion*, the **field-strength tensor** $F_{\mu\nu}$ is a two-index tensor, given by

$$F_{\mu\nu} = \partial_\mu A_\nu - \partial_\nu A_\mu. \tag{9.8}$$

Notice that the indices μ and ν have switched between the two terms here—if they hadn't, we would just get 0. Unlike the metric tensor, which is symmetric ($g_{\mu\nu} = g_{\nu\mu}$), the field strength tensor is therefore **antisymmetric**: $F_{\mu\nu} = -F_{\nu\mu}$.

More important, the field-strength tensor is **gauge invariant**—when we do a gauge transformation, $F_{\mu\nu}$ doesn't change at all. We can see this if we take the gauge transformation of A_μ from (9.7) and plug it into the definition of the field strength (9.8), obtaining

$$\partial_\mu \partial_\nu \theta - \partial_\nu \partial_\mu \theta = 0. \tag{9.9}$$

This is automatically zero because partial derivatives always commute with each other: we get the same answer no matter which order we take them in. Whenever we're working with gauge theories, a big part of the game is to make sure that physically relevant quantities are gauge invariant. The field-strength tensor $F_{\mu\nu}$ is a paradigmatic example: (9.9) shows that it doesn't change when we do (9.7) to A_μ. The gauge field itself is not gauge invariant but its corresponding field-strength tensor is.

The relationship of $F_{\mu\nu}$ to the electric and magnetic fields $\vec{E}$ and $\vec{B}$ is that they are the same thing. More specifically, if we think of $F_{\mu\nu}$ as a 4 × 4 matrix, the electric and magnetic fields are its components:

GAUGE THEORY

$$F_{\mu\nu} = \begin{pmatrix} 0 & E_x & E_y & E_z \\ -E_x & 0 & -B_z & B_y \\ -E_y & B_z & 0 & -B_x \\ -E_z & -B_y & B_x & 0 \end{pmatrix}. \tag{9.10}$$

So $F_{01} = E_x$, $F_{23} = B_x$, and so on. That's why A_μ is sometimes called the vector "potential"—just like the gravitational force in Newtonian mechanics can be thought of as the derivative of a gravitational potential field, the electric and magnetic fields (which directly give rise to forces) can be thought of as appropriate derivatives of A_μ. The field strength tensor is directly analogous to the Riemann tensor in general relativity—they are both invariant measures of the "curvature" of the underlying potentials.

You will sometimes hear that Maxwell himself didn't know about gauge invariance, because he worked directly with the electric and magnetic fields. That's not at all true—Maxwell knew that you could define a vector potential and derive the electric and magnetic fields from it, and that the potential could be gauge-transformed (though he didn't call it that) without altering $\vec{E}$ or $\vec{B}$. He didn't know about special relativity, of course, so he never organized the electromagnetic fields into a spacetime tensor. Nor did he know about the electron field; he didn't even know about electrons. It's in the context of quantum field theory that the gauge field A_μ and its properties really become indispensable.

THE ELECTRON AND THE POSITRON

Let's put this together to do some physics.

If you remember Chapter 5, Interactions, you know that by "doing physics" we mean "writing down a Lagrangian, and from that constructing Feynman diagrams to represent interactions." Gauge invariance is going to be crucial to this process. The game we're going to play is to try to construct something similar to our previous Lagrangian for a scalar field, but now with the extra requirement that the theory be invariant under gauge symmetry.

The action is the integral over all spacetime of the Lagrange density, $S = \int \mathcal{L} \, dt \, d^3x$, and the Lagrange density—usually just called "the Lagrangian" for short—is constructed from the fields and their derivatives. The Lagrangian for fields will be a sum of different terms. There are kinetic and gradient terms with derivatives of the fields with respect to space and time, usually just unified into a "kinetic term" since space and time are unified in relativity. There could also be mass terms of the form (field)^2, and interaction terms involving more than two powers of the fields.

$$\mathcal{L} = (\text{kinetic term}) + (\text{mass term}) + (\text{interactions}). \quad (9.11)$$

We're going to want to make sure the Lagrangian as a whole is gauge invariant. Individual terms don't need to be, if they transform in a way that is compensated by the change in some other term.

Let's think about the electron field first. It has a kinetic term, first written down by Paul Dirac, but let's not worry about it for the moment. There are a number of mathematical complications stemming from the fact that the electron is spin-½, and those led Dirac to posit his famous "Dirac matrices," which we'll largely be sidestepping.

Consider the mass term. You might guess that if the electron field is ψ_e, its mass term should simply be $(\psi_e)^2$. But that can't be right. Look what happens under a gauge transformation like (9.3):

$$\left(\psi_e\right)^2 \rightarrow \left(e^{i\theta}\psi_e\right)^2 = e^{2i\theta}\psi_e^2. \quad (9.12)$$

GAUGE THEORY

That's not invariant; it picks up a phase factor $e^{2i\theta}$. Does this mean that electrons can't have mass after all?

Clearly not, since electrons do have mass (0.511 MeV). We just have to think more carefully about how to make a gauge-invariant mass term. Happily we have one other arrow in our quiver: because the electron field is complex, we can also work with the complex-conjugated field ψ_e^*. We know what happens to it under a gauge transformation, since we know what happens to ψ_e itself and complex conjugation changes every i to $-i$:

$$\psi_e^* \rightarrow \left(e^{i\theta}\psi_e\right)^* = e^{-i\theta}\psi_e^*. \tag{9.13}$$

This allows us to make something gauge invariant, namely $\psi_e^* \psi_e$. In fact, we know right away that this is gauge invariant, since it's simply the modulus squared $|\psi_e|^2$, and the whole point of a U(1) rotation is that it leaves the modulus invariant while shifting the phase of the complex field. But just to be sure, we can also work it out:

$$\psi_e^*\psi_e \rightarrow \left(e^{-i\theta}\psi_e^*\right)\left(e^{i\theta}\psi_e\right) = e^{-i\theta+i\theta}\psi_e^*\psi_e = \psi_e^*\psi_e. \tag{9.14}$$

A U(1) gauge transformation leaves this combination completely unchanged, just as we had hoped. That's what we're looking for in a well-behaved Lagrangian.

Are we really sure that it's okay to use the complex conjugate of the original field to construct our mass term? The answer is yes, but it took a while for physicists to wrap their heads around why. The trick is to think of ψ_e^* as a field in its own right, in addition to ψ_e rather than being derived from it. And if that's true, there will be quanta of ψ_e^* that we will see as particles.

Indeed, and that particle is the **positron**, the antiparticle of the electron. The reasoning we've gone through is a simplified, toy version of how Dirac derived the Dirac equation for the electron back in 1928. His equation seemed to imply that there must be another particle with the same mass as the electron but an opposite electric charge. You may have heard that this prediction had something to do with understanding the spin of the electron, and historically that did play a role. But in fact any charged particle—that is, one with a U(1) gauge symmetry—is going to have an antiparticle, no matter what its spin might be.

The mass term in the Lagrangian that fixes the mass of the electron also does the same for the positron, so they have no choice but to have the same mass. Physicists were reluctant to accept the prediction of a new positively charged electron-like particle—why hadn't they discovered it already?—but the positron was ultimately found by Carl Anderson in 1932.

You may have heard that the Higgs boson field is responsible for particles getting mass. If so, why were we able to write down a mass term for the electron without ever talking about the Higgs? The answer is that the need for the Higgs in the Standard Model of particle physics is quite special. It has to do with the fact that the weak interactions violate parity—they treat left-handed-spinning particles differently from right-handed-spinning particles. Because Dirac wasn't thinking about parity or the weak interactions, his theory had no trouble just including the mass term for the electron and getting on with its life.

Meanwhile, the symmetry of gauge invariance is enough to forbid any mass for another famous particle: the photon.

LONG-RANGE FORCES

So let's think about the Lagrangian for the photon field A_μ. Once again, gauge invariance is going to play a major role. For the kinetic term, we already know an easy way to make a gauge-invariant tensor: the field strength $F_{\mu\nu}$, defined in (9.8). But the Lagrangian is

GAUGE THEORY

supposed to be a scalar (zero indices), not a tensor with two indices. Remember that the field strength is analogous to the curvature tensor in general relativity. In that case we were able to construct a scalar R that could be used to make a Lagrangian, but that doesn't work for electromagnetism. What we can do is multiply the field strength by itself, to make $(F_{\mu\nu})^2$. And indeed, that is the correct kinetic term for electromagnetism. Gauge invariance makes our life easier.

Quick procedural note: Whenever we multiply tensors by each other, we need to "absorb the indices" by summing over their possible values in order to make a scalar quantity. This can be done by contracting with appropriate factors of the metric $g_{\mu\nu}$ and inverse metric $g^{\mu\nu}$, using those to raise and lower indices and making sure to sum over repeated indices when one is up and one is down. In this book we'll completely ignore that subtlety. We're going to write $(A_\mu)^2$ when we really mean $A_\mu A^\mu$, and likewise $(F_{\mu\nu})^2$ for $F_{\mu\nu} F^{\mu\nu}$, and so on. Whenever you see something like that going forward, assume that we are implicitly raising and lowering indices and summing over them appropriately.

So what about a mass term for the photon? The straightforward thing to imagine would be something like $(A_\mu)^2$. But that is blatantly not gauge invariant, as we can check by plugging in (9.7):

$$\text{U}(1): \left(A_\mu\right)^2 \to \left(A_\mu + \partial_\mu \theta\right)^2 = \left(A_\mu\right)^2 + 2A_\mu(\partial_\mu \theta) + \left(\partial_\mu \theta\right)^2. \quad (9.15)$$

This term clearly changes when we do a gauge transformation (at least whenever $\partial_\mu \theta$ is not zero), so it's not gauge invariant. For the electron we could fix this problem by invoking the complex-conjugate field. But A_μ is a real-valued vector field; it doesn't have a separate complex conjugate. What are we to do?

The answer is: give up. Without introducing entirely new fields (analogous to the Higgs) that would drastically change the physics of electromagnetism, there is no way to give mass to the photon in a gauge-invariant way. Thus we see another crucial physical implication of gauge invariance: the force-carrying particles are naturally massless.

Similar reasoning applies to gravity as described by general relativity. In both cases we get long-range forces carried by massless particles because there are symmetries prohibiting the particles from obtaining mass, leading to inverse-square force laws (Coulomb's law for electromagnetism, Newton's law of universal gravitation for gravity). That's why it's those two forces that are most evident in our macroscopic, human-scale lives. The energy of a particle is related to its momentum and mass by $E^2 = p^2 + m^2$, so the minimum energy a particle can have (namely, when $p = 0$) is $E = m$. It takes a certain minimum energy to create a massive particle; in terms of the forces they transmit, that implies that those forces diminish rapidly with distance. Massless particles, meanwhile, can be as low-energy as we wish. It takes little effort for low-energy virtual photons or gravitons to journey across vast distances of space, making long-range forces possible.

QED INTERACTIONS

So we have the basic ingredients of QED: the complex electron field, its complex conjugate the positron field, and the gauge connection field that gives us electromagnetism and photons. The final step in assembling the theory is specifying how photons and electrons/positrons interact with each other. Back in equation (5.6) we wrote it as a simple product of the three fields (electron, positron, and photon). But in those simpler times we didn't know about gauge invariance, or that the positron field is the complex conjugate of the electron field. We were basically on the right track, but now we have to be a bit more careful.

We want our interaction to be gauge invariant, keeping in mind the transformation properties of the electron (9.3) and the photon (9.7). We can cancel the $e^{i\theta}$ from the electron field ψ_e by always using it in multiplicative combination with the positron field ψ_e^*. But that

GAUGE THEORY

photon looks like it's going to give us trouble. A gauge transformation adds $\partial_\mu \theta$ to A_μ, and it's not clear what that can cancel out.

Wait a minute, maybe it is clear. Remember (9.6), the gauge-invariant version of "keep the electron field constant"? The expression on the left-hand side just picks up an overall phase $e^{i\theta}$ when we do a gauge transformation. There are ugly additional pieces that would enter for either of the terms individually, but when put together those all cancel out. Perhaps we can take advantage of that.

To simplify our notation and keep our minds on the essential physics, let's hide some indices, writing ∂ for ∂_μ and A for A_μ. Then the left-hand side of (9.6) is $\partial \psi_e - iA\psi_e$. We're not actually setting this equal to 0—we're just taking advantage of the fact that the only thing that happens to it under a gauge transformation is that it picks up an overall phase $e^{i\theta}$. There is an obvious way to get rid of even that phase: multiply by the complex-conjugate (positron) field, which transforms with the opposite phase, as $\psi_e^* \to e^{-i\theta} \psi_e^*$. Thus, we can construct a completely gauge-invariant combination that is eligible to appear in the Lagrangian:

$$\psi_e^* \left(\partial \psi_e - iA\psi_e \right) = \psi_e^* \partial \psi_e - i\psi_e^* \psi_e A. \tag{9.16}$$

Intriguing! The second term looks basically like the interaction we wrote down in Chapter 5—a product of the electron, positron, and photon fields, in slightly different notation. But what is that first term, with just the positron and a derivative of the electron? Easy: that's the kinetic term. Unlike for scalar fields or for the photon field, the kinetic term for electrons (and other spin-½ particles) requires just a single derivative rather than two derivatives.

Another secret is thereby revealed: the form of the interaction between photons and electrons/positrons in QED can be thought of as the result of starting with the kinetic term, demanding it be gauge invariant, and realizing that we have to compensate in an appropriate way by including a term with the connection, aka the gauge field. The result differs from (5.6) by an overall numerical factor, but that's just because we've been pretty casual about overall numerical factors. The second term in (9.16) truly is the QED interaction Lagrangian, completely determined by the requirements of gauge invariance. That nice feature will hold true for any gauge theory, not just QED.

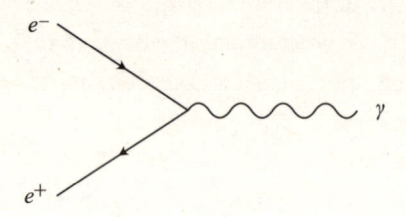

We can also see, behind the scenes, Noether's theorem at work. The theorem tells us that continuous symmetries imply conserved quantities. The U(1) symmetry of electromagnetism certainly counts, and the associated quantity is electric charge. And now we see how gauge invariance makes it happen. In order to get a gauge-invariant Lagrangian, we needed to include the positron field as well as the electron field in the interaction term $\psi_e^* \psi_e A$. Which means that the corresponding Feynman diagram vertex will have both an electron and a positron coming in, with a photon going out. One -1 charge and one $+1$ charge coming in, one 0 charge going out, conserving charge overall. We can never have a vertex with two electrons coming in and a single photon going out; gauge invariance won't allow it.

TEN

PHASES

Quantum electrodynamics has a fascinating and storied history. But in some sense its success was not surprising. We know that quantum mechanics works, and we know that there is electromagnetism, so they had to come together somehow. There was some initial puzzlement about renormalization, but happily everything worked out.

The real triumph of quantum field theory, which didn't seem at all inevitable at the time, was in accounting for the other forces of nature, in particular the strong and weak nuclear forces. The idea that these could be some kind of generalizations of QED, this time based on non-abelian (non-commuting) gauge symmetries, was an attractive one that started with Chen Ning Yang and Robert Mills in 1954, so such models are known as **Yang-Mills theories**. But there were a number of immediate obstacles, and more than a few physicists became skeptical that the program would ever work. Its eventual success came from a better understanding of the richness of non-abelian gauge theory, in particular the fact that theories can manifest in different **phases**.

The word "phase" is borrowed from macroscopic physics, where it refers to the way that a single kind of matter can appear in different states, with correspondingly different physical properties. Water, for example, can appear in the form of a solid (ice), liquid, or gaseous vapor, depending on its temperature and pressure. The different phases of a material will have different characteristics, like density or the speed of sound inside the material, despite being made of the same underlying stuff.

Phases of gauge theories work analogously: a similar set of underlying ingredients can exhibit different kinds of observed behavior. Theories like electromagnetism and general relativity are said to be in the **Coulomb phase**, characterized by long-range forces obeying an inverse-square law. In particle language we would say that the gauge bosons are massless and weakly interacting, which is true for both photons and gravitons. A theory could be in the **confined phase** if the gauge bosons are still massless but they are strongly interacting and therefore captured inside composite particles. That's the case for the gluons in the strong interactions (QCD). And finally we have the **Higgs phase**, where the underlying gauge symmetry is spontaneously broken. Then the gauge bosons become massive and the force becomes short-ranged, as in the weak interactions. This incredible richness of structure was largely unanticipated by the pioneers of gauge theory; the equations, as usual, are smarter than we are.

QCD

In the previous chapter we sang the praises of gauge invariance, an enormously helpful tool when it comes to understanding the dynamics and interactions of quantum fields. One of the big implications is that the gauge bosons, such as photons and gravitons, must be massless, which is why they manifest as long-range forces.

Wait a minute. We have already foreshadowed that the other forces of nature, the strong and weak nuclear forces, are also described by gauge theories. But those forces are most definitely *not* long-range. A typical range associated with the strong force would be the Compton wavelength of a proton. So look up the mass of the proton in electron volts, convert to inverse centimeters, and take the reciprocal to get a distance. We end up with a number of order 10^{-14} cm. For the weak force, do the same thing but with the W boson instead of the proton, to get 10^{-16} cm. Very short ranges indeed. What's up?

In both cases of the strong force and the weak force, there are subtle and fascinating reasons why the forces are short-range even though they are based on gauge symmetries. And because the universe likes to keep things interesting, the reasons are completely different in the two cases.

Let's think about the strong nuclear force first: quantum chromodynamics, or QCD, which is responsible for binding quarks and gluons into composite particles like baryons and mesons. (The force holding protons and neutrons together in nuclei can be thought of as spillover from what's going on inside the nucleons themselves.) There are two features that make QCD so different in practice from QED: gluons interact directly with other gluons, and renormalization makes the interaction stronger, rather than weaker, at lower energies.

QCD is based on an SU(3) symmetry from rotating quark fields in three-dimensional complex color space. It has a long and somewhat tortuous history. The idea of generalizing from abelian to non-abelian gauge symmetries came from Yang and Mills in 1954, but they didn't know about quarks. Instead they were trying to work directly with protons and neutrons, and none of the attempts really panned out. It wasn't until 1964 that Murray Gell-Mann and George Zweig independently proposed the idea of quarks. Soon thereafter, Otto Greenberg and separately Moo-Young Han and Yoichiro Nambu suggested SU(3) gauge theories with quarks as the fundamental constituents, but the details of their models were very different from the current picture of QCD. It was Gell-Mann and Harald Fritzsch who introduced color charges as we now understand them in 1971, and in separate papers with William Bardeen and Heinrich Leutwyler they put together the theory as we now know it (without yet appreciating some of its crucial properties). Gell-Mann, who had a genius for terminology, dubbed the theory "quantum chromodynamics" by analogy with quantum electrodynamics.

In many ways the basic structure of QCD is much like that of QED. In both cases we have a connection gauge field, which gives rise to photons in QED and gluons in QCD. In QED we have electric charge, which is just a number, while in QCD we have color charges that live in a three-dimensional red/green/blue vector space. In both cases the force-carrying particles are massless, ultimately because of restrictions imposed by gauge invariance. The form of the kinetic terms and the interactions are largely the same.

There is one important difference: SU(3) is a non-abelian group, whereas U(1) is an abelian group. A U(1) transformation is just rotation in the complex plane, or equivalently multiplication by a phase factor, $R = e^{i\theta}$. So it doesn't matter what order we do them in: $R_1 R_2 = e^{i(\theta_1 + \theta_2)} = R_2 R_1$. But an SU(3) transformation M is a 3×3 matrix, and the order in which we do two of them matters:

$$M_1 M_2 \neq M_2 M_1. \tag{10.1}$$

That seems like a mathematical nicety, but it has enormous physical consequences. The fact that U(1) is abelian means that successive transformations "pass right through each other," and through a bit of math (which we are hiding) this translates directly to the fact that physical photons pass right by without interacting. SU(3) transformations do not pass right through each other, and gluons, likewise, tend to bump into each other.

More formally, at the level of direct interactions—fundamental Feynman diagram vertices that we would read off from the basic Lagrangian—photons interact only with electrically charged particles. And the photon itself is neutral, so photons don't interact directly with themselves. (There are effective four-photon interactions once we impose a cutoff, but those are generally pretty weak, and anyway we know they are really induced by loop diagrams with charged particles.) By contrast, gluons carry a form of color charge themselves. Essentially each gluon carries both a color (red, green, or blue) and an anti-color. So a red quark and a green quark can interact by exchanging a red/anti-green gluon, thereby changing their own colors in the process.*

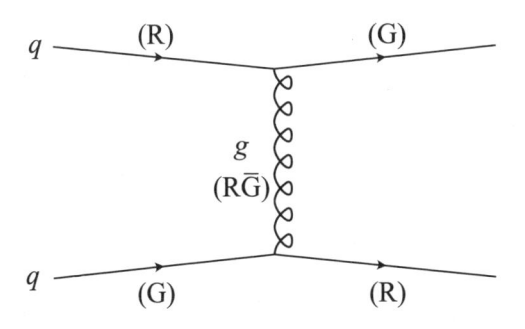

This gluon-exchange diagram is not so different from the QED case, all by itself. But there are also both three-gluon and four-gluon vertices as part of the fundamental ingredients in QCD Feynman diagrams. The existence of these vertices is a consequence of the non-abelian nature of the SU(3) gauge symmetry.

---

* This quick story might lead you to think there should be nine different kinds of gluons (three colors times three anti-colors), but one overall combination doesn't contribute, so there are actually only eight types of gluons. That's consistent with SU(3) being an eight-dimensional group ($3^2 - 1 = 8$). There is always one kind of gauge boson per dimension of the gauge group.

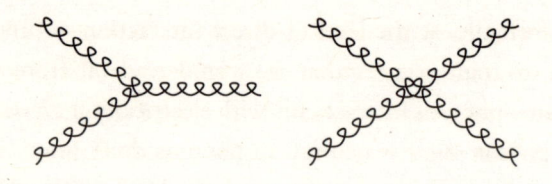

CONFINEMENT

The other ingredient that plays a defining role in the strong interactions comes from our discussion in Chapter 6 of the renormalization group and running coupling constants. There we explained that the effective fine-structure constant of QED gets larger as we consider higher and higher energies. It wasn't until the 1970s that physicists realized, to their surprise, that in certain non-abelian gauge theories the coupling runs in the opposite way: it appears larger at low energies, smaller at high energies. The calculation was worked out by David Politzer, David Gross, and Frank Wilczek in 1973, for which they shared the 2004 Nobel Prize in Physics. This property is called **asymptotic freedom**, since as interaction energies get asymptotically larger, the coupling goes to 0—the quarks and gluons act like free particles. But on the flip side, as the interaction energies get smaller—which, remember, corresponds to larger distances—the coupling gets larger. In principle it could get infinitely large.

In practice, what this means is that colored quarks and gluons never actually get to large distances and low interaction energies. Rather, we have the phenomenon of **confinement**: colored particles are always bound up with other colored particles in colorless combinations. Confinement explains how gluons can be massless, yet the strong nuclear force has such a short range: gluons move at the speed of light, but they don't travel very far. They keep interacting with quarks and other gluons; if they threaten to wander off too far from their companions, they are irresistibly pulled back.

Of course, you know by now that the previous paragraph is just an evocative story. What's really going on is that there are quantum fields interacting with each other. In QCD, the basic fields we start with correspond to quarks and gluons, but those are not the individual particle-like states we end up with. Rather, the quark and gluon fields settle into certain collective particles that we call **hadrons**. The quarks and gluons inside a hadron aren't literal point-like particles constantly moving around and bumping into each other; generally the fields are completely stationary, or nearly so. But there's nothing wrong with telling an evocative story as long as we understand its limitations.

You might think, "Well, what if I try to pull quarks apart from each other? What goes wrong?" This leads to another evocative story. It starts from recognizing that the strongly interacting gluon field around a quark doesn't spread out evenly in all directions like the electric field from a charged particle. Rather, it gets confined into a **flux tube** that extends to some other quark. Consider an up quark and an anti-up quark, which can combine together to form (at least for a while) a particle called the **neutral pion**. If we took really tiny tweezers and tried to pull the two quarks apart from each other, we'd be stretching out the flux tube. That requires energy, since the tube has a constant amount of energy per unit length. Eventually we've put so much energy into the system that it's favorable to just make another quark/anti-quark pair. This pair can spontaneously appear along the flux tube, which then splits into two flux tubes. It's much like ends of string: if you cut a string that initially has two ends, you don't get two separated ends, you get two pieces of string with two ends each.

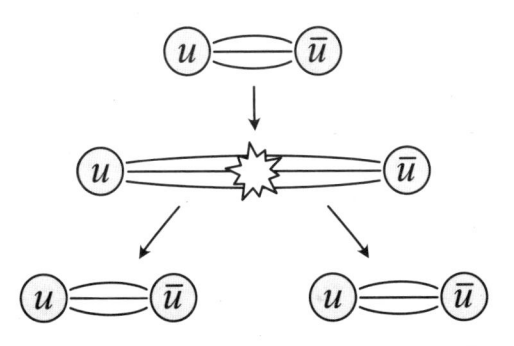

Before we knew about gluons or strongly interacting flux tubes, it was noticed that certain hadrons fell into patterns we might expect from the vibrational modes of a string. This led to the invention of something called **string theory**. The idea never really fit as a theory of just strong interactions, but it grew into a framework for quantum gravity as well as all the other interactions together, and it is still extremely popular among theoretical physicists.

SYMMETRY BREAKING

That just about does it for the electromagnetic and strong interactions. Both forces are based on gauge symmetries, and the corresponding gauge bosons are massless. A similar story holds for gravitation, which is described by general relativity. Gravitons interact with each other, but only very weakly, so gravitons are not confined, and gravity is in the Coulomb phase.

Finally we turn to the weak interactions, where there are three gauge bosons: the charged W^+ and W^- and the neutral Z^0. They are not massless: the W's are about 80 GeV, and the Z is about 91 GeV. How can the force-carrying bosons be massive if gauge invariance prohibits us from writing down something like $m^2(A_\mu)^2$ in the Lagrangian? The answer is because the underlying symmetry is spontaneously broken.

To keep things simple, let's consider a toy model with an SO(2) symmetry group. As a group, SO(2) is the same as U(1), but we are going to think of it as rotations in a real two-dimensional vector space, rather than a single complex dimension. Say we have a scalar field Φ that is a vector in this two-dimensional internal space. That means it has components Φ_1 and Φ_2, and the whole field can be written as

$$\Phi = \Phi_1 \vec{e}_1 + \Phi_2 \vec{e}_2, \tag{10.2}$$

PHASES

where $\vec{e}_1$ and $\vec{e}_2$ are basis vectors. The SO(2) symmetry rotates the vector Φ within this space.

To make a Lagrangian, we need to construct quantities that are invariant under the symmetry. There is an obvious example, the field squared:

$$|\Phi|^2 = (\Phi_1)^2 + (\Phi_2)^2. \tag{10.3}$$

This is just the length-squared of the vector Φ, which is left unchanged when we rotate by any angle around the origin. So it's easy to make an SO(2)-invariant mass term, $m_\Phi^2 |\Phi|^2$. Likewise, we can make an interaction term just by squaring again to get $|\Phi|^4$. Since $|\Phi|^2$ is invariant under the symmetry, any function of it will be as well.

Let's consider the following potential for Φ:

$$V(\Phi) = -\mu^2 |\Phi|^2 + \lambda |\Phi|^4. \tag{10.4}$$

This is known as a "sombrero" or **Mexican-hat potential**, for reasons that become clear when we plot it, as we have on the next page in a cutaway view. The parameter μ has dimensions of mass, and λ is dimensionless (as we can figure out, because a scalar field such as Φ has dimensions of mass, and the potential is part of the Lagrange density, which has dimensions of $[E]^4$, which is the same as $[M]^4$).

The interesting thing is that we have a minus sign out in front of the $\mu^2 |\Phi|^2$. We didn't need to—the sign could be either plus or minus, and we're just exploring this possibility. As a result, the minimum of the potential is not the origin at $\Phi = 0$. Near the origin, the $|\Phi|^2$ becomes important before the $|\Phi|^4$ term, and that term has a negative coefficient. So the potential goes down as we initially go away from the origin, then eventually goes up again once the $|\Phi|^4$ term starts to dominate.

$V(\Phi)$, Φ_2, $\overline{\Phi}$, Φ_1

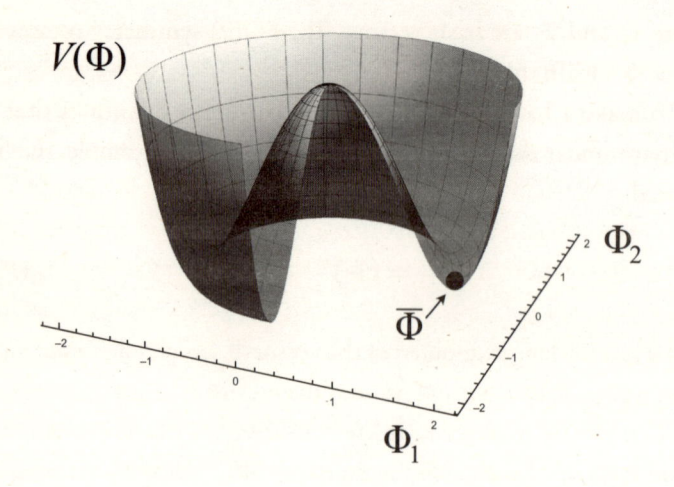

There is nothing weird or wrong about such behavior. There is no rule that says the minimum of the potential has to be at the origin. There is a rule that says there needs to *be* a minimum somewhere. Otherwise, there is no stable vacuum state, and the field rolls down the potential forever. That might be an interesting alternative universe, but it doesn't seem to be the world in which we live.

Since we usually think of ourselves as living near the vacuum (minimum-energy) state, the field is not going to be centered around 0; it will fall down into the brim of the hat. The Mexican-hat potential is rotationally symmetric about the origin, reflecting the underlying $SO(2)$ symmetry. So we can't say ahead of time where the field will fall—all the points in the brim are equivalent to each other. But it will fall somewhere. We label the value where it falls $\overline{\Phi}$, called the **vacuum expectation value** of the field.* ("Expectation value" rather than simply "value" because we are doing quantum mechanics, after

---

* You might wonder whether the field could fall to different values at different points in space. This is tricky because of the gauge symmetry, but essentially yes, that can happen. Overall that's a higher-energy state than a constant field, so what happens is that the field straightens itself out as that energy converts into other particles.

all. If you were to repeatedly measure the value of the field, you would get slightly different answers, but they would average out to be $\bar{\Phi}$.)

All of the underlying physics of this scalar field is completely invariant with respect to SO(2) transformations. But the specific place $\bar{\Phi}$ where the field ends up is *not* invariant; a rotation moves it along the brim of the hat. By randomly falling to some particular point along the brim of the hat, the field value is now violating the underlying symmetry. That's why this phenomenon is called **spontaneous symmetry breaking**—because the underlying physics is invariant, but the specific vacuum expectation value is not.

Everything we have thus far said about spontaneous symmetry breaking would apply equally well if the symmetry were global (you can only do a uniform rotation throughout spacetime) rather than gauge (you can specify different rotations at each point). The global case was studied first in the 1960s by Yoichiro Nambu, Jeffrey Goldstone, and others. Something interesting happens in that case: because of the underlying symmetry, there is always a direction in field space where we can start at the vacuum-expectation value and move such that the potential remains constant. (In our picture, it's the brim of the hat.) A constant potential, even just along certain directions in field space, corresponds to a particle with zero mass, called a **Goldstone boson** (or Nambu-Goldstone boson). There turns out to be a theorem one can prove that whenever you spontaneously break an exact global symmetry, you are left with one or more massless scalar particles. That's amusing, but also problematic if you're trying to describe the real world, since no such particles are known to exist.

Happily, spontaneously breaking a gauge symmetry is an entirely different kettle of fish. Let's consider that case.

THE HIGGS MECHANISM

In the 1960s, physicists hadn't yet figured out asymptotic freedom and confinement, but they knew perfectly well that the strong and weak nuclear forces were short-range. They were on the search for a way to reconcile the Yang-Mills proposal of a non-abelian gauge theory with the problem of the gauge bosons being massless. It seemed promising to consider spontaneous symmetry breaking, which had been shown to play a crucial role in understanding superconductivity. But there was the problem of the massless Goldstone bosons, which no one had ever observed.

The fact that spontaneous symmetry breaking works quite differently for gauge symmetries as compared with global symmetries was first pointed out by condensed-matter physicist Philip Anderson and deployed in a particle-physics context by three groups in 1964: Robert Brout and François Englert; Peter Higgs; and Gerald Guralnik, Carl Hagen, and Tom Kibble. The basic idea is that two problematic features—massless gauge bosons and massless Goldstone bosons—cancel each other out. When a gauge symmetry is spontaneously broken, the associated gauge bosons "eat" the Goldstone bosons. The Goldstones disappear as independent particles, and the gauge bosons become massive rather than massless. It is convenient to denote what happens as simply the Higgs mechanism, but all these folks deserve some credit. Englert and Higgs shared the Nobel Prize in 2013; Brout had already passed away, and the Nobel is not awarded posthumously. (That's almost half a century between proposing an idea and winning the Nobel for it, if you're counting.)

It's now time to take the hard work we did to understand how to make gauge-invariant Lagrangians and use it to see explicitly how the gauge bosons become massive. The trick lies in the kinetic Lagrangian for the scalar field that is going to get a vacuum expectation value in a Mexican-hat potential. Let's consider our SO(2) example with scalar Φ, and now we're explicitly considering a gauge symmetry, so we also have a connection field A_μ.

In an ordinary scalar field theory, the kinetic term would look something like $(\partial_\mu \Phi)^2$, but by itself that's not gauge invariant. Instead we hark back to (9.6), where we made a gauge-invariant combination of the derivative of the electron field and the connection times that field. Now we are thinking of our scalar Φ rather than the electron, and we're thinking about SO(2) rather than U(1) so there are no imaginary numbers, but the basic ideas are the same. A candidate gauge-invariant kinetic term is*

$$|\partial_\mu \Phi - A_\mu \Phi|^2 = (\partial_\mu \Phi)^2 + (A_\mu)^2 |\Phi|^2. \quad (10.5)$$

It looks like, on the right-hand side, the first piece is an ordinary kinetic term and the second is an interaction involving two gauge bosons and two Φ particles. And indeed it would be, if it weren't for spontaneous symmetry breaking.

When we talk about "particles" in quantum field theory, we're thinking of quanta arising as excitations in the field, small perturbations above the vacuum state. But the vacuum state here isn't $\Phi = 0$, it's $\Phi = \overline{\Phi}$. So to reveal what kinds of particles this theory predicts, it is helpful to change variables:

$$\Phi(x, t) = \overline{\Phi} + \phi(x, t). \quad (10.6)$$

Here $\overline{\Phi}$ is the vacuum-expectation value, which is just a constant, and $\phi(x, t)$ is a field that tells us how far away we are from $\overline{\Phi}$. We don't have to make this substitution, it's just a convenient thing to do, since now the vacuum value of our field is $\phi = 0$. It's the shifted field ϕ, not the original field Φ, whose excitations are naturally interpreted as particles.

---

* We're hiding some mathematical trickery that accounts for why we don't have a cross term $-2A_\mu \Phi (\partial_\mu \Phi)$, but trust me this is okay. To dig further, we would have to pay closer attention to the fact that Φ is a two-component vector and each component of A_μ is a 2×2 matrix.

Let's see what happens when we plug the change of variables (10.6) into the second "interaction" part of the kinetic term (10.4):

$$\left(A_\mu\right)^2 |\Phi|^2 = \bar{\Phi}^2 \left(A_\mu\right)^2 + 2\bar{\Phi}\phi\left(A_\mu\right)^2 + \phi^2\left(A_\mu\right)^2. \quad (10.7)$$

The last term here is just the interaction we previously discussed, between two scalars and two gauge bosons. To interpret the second-to-last term, remember that $\bar{\Phi}$ is just a constant number, not a dynamical field. So this term is another interaction, between a single ϕ and two gauge bosons, with $2\bar{\Phi}$ acting as a coupling constant.

But the first term on the right-hand side looms with implication. $\bar{\Phi}$ is just a constant, so the only field content there is $(A_\mu)^2$. So really this is a mass term for the gauge boson! Gauge invariance was supposed to prevent that from ever happening. What's going on?

In one legitimate sense, what's going on is the mathematical manipulations we just went through. In a gauge theory, we cannot simply plop down a mass term for the gauge field into our Lagrangian without violating gauge invariance. But spontaneous symmetry breaking brings an effective mass term into existence, with the role of the mass being played by the vacuum expectation value of the symmetry-breaking field. (If we had been more careful, there would also be coupling constants relating $\bar{\Phi}$ to the mass of A_μ, but we're going for general concepts here rather than precise formulas.) Although we can't see it explicitly from what we've done, the massless Goldstone boson has also disappeared, having been eaten by the gauge boson.

But you would surely be disappointed if there were not an evocative story to be told. Here's one way of thinking about it: a massless, non-asymptotically free gauge field like we have in electromagnetism or gravity gives rise to a long-range, inverse-square force law, basically because lines of force travel forever and they dilute away with distance, as discussed in *Space, Time, and Motion*. Now we have a new field Φ pervading space, with a nonzero expectation value everywhere.

And it's a charged field, as far as our gauge bosons are concerned; it transforms under the gauge transformation. So the lines of force don't travel freely to infinity. They travel through the ambient scalar field and are gradually absorbed by it, like rays of light fading quickly in a smoggy atmosphere. That's why this gauge field is short-range rather than long-range. And that, in turn, corresponds to massive gauge bosons rather than massless ones.

ELECTROWEAK THEORY

The pioneers of the Higgs mechanism were targeting the strong nuclear force, but the idea eventually found a home in our modern understanding of the weak nuclear force. The road to getting there was, characteristically for the history of quantum field theory, twisty and unpredictable.

The first pretty-successful theory of the weak interactions was Enrico Fermi's theory from 1933 of beta decay, which described the decay of a neutron into a proton, an electron, and an anti-neutrino. It was an influential early example of a quantum field theory and featured a place for the neutrino, which at the time was still purely hypothetical. It was also non-renormalizable, which was concerning to physicists at the time. (Fermi posited a direct interaction between four fermions, the fields of which have dimension $[E]^{3/2}$, so his interaction term has dimension $[E]^6$, and is therefore an irrelevant operator.) These days we would say that Fermi's theory is a perfectly respectable low-energy effective theory for weak interactions.

One way to possibly construct a renormalizable theory that reduces to Fermi's at low energies might be to introduce a charged, massive boson that could mediate the interaction, replacing a direct vertex with four fermions with two vertices, each with two fermions and the new boson. In the modern language of quarks and Feynman diagrams, a neutron could decay to a proton when one of its down quarks (charge $-\frac{1}{3}$) could convert to an up quark (charge $+\frac{2}{3}$) by emitting a

W boson with charge -1, which would subsequently convert into an electron and an anti-neutrino.

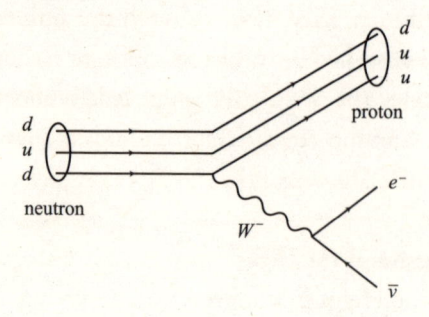

If you have been wondering how we distinguish between the electron neutrino, muon neutrino, and tau neutrino, this is how: when the W converts into a charged lepton and an anti-neutrino, it's always the type of anti-neutrino that is associated with the lepton that was produced.

But since the Higgs mechanism wasn't yet known, it was hard to construct a theory with a massive gauge boson. In the 1950s Julian Schwinger made an attempt, using a model with an SU(2) symmetry, but his theory had the unwanted consequence of also predicting a massless neutral gauge boson. Eventually he had the idea of converting that bug into a feature, interpreting the massless neutral boson as the well-known photon. From this was born the ambition of **unifying** the weak and electromagnetic interactions. Schwinger handed off the problem to his talented student Sheldon Glashow, who suggested a model with both an SU(2) and a U(1) symmetry. Over in the UK, Abdus Salam and John Ward explored similar models. But the mechanism by which any of the gauge bosons were supposed to become massive was still a mystery at the time, in the early 1960s. It was Steven Weinberg, and also Salam in unpublished work, who introduced the Higgs mechanism of spontaneous gauge symmetry breaking in attempts to unify electromagnetism and the weak force, leading to what we now call the **electroweak theory**.

The resulting theory is an intricate construction. We start with both an SU(2) gauge symmetry (a three-dimensional group, so three gauge bosons) and a separate U(1) gauge symmetry (one gauge boson), as well as a complex scalar field that transforms as a two-dimensional vector under the SU(2), and is also charged under the U(1). When this field—now known simply as the Higgs field—gets a vacuum expectation value, both the SU(2) and the U(1) symmetries are spontaneously broken, but a certain combination of them remains nevertheless unbroken and persists as the U(1) of electromagnetism. There are three massive gauge bosons—the charged W^+ and W^-, and the neutral Z^0—as well as the massless photon. So even though the theory starts as SU(2) × U(1) and is broken down to just U(1), the final gauge symmetry is not precisely the original U(1) part but a combination of that and part of the SU(2).

I told you it was intricate. But there is one other wrinkle that turns out to be tremendously important: in the 1950s, following a suggestion by theorists Tsung-Dao Lee and Chen Ning Yang, experimental physicist Chien-Shiung Wu demonstrated that the weak interactions violate **parity**, the transformation that exchanges a physical configuration for its mirror image. In the Weinberg-Salam electroweak theory, fermions like electrons and neutrinos and quarks all start out as perfectly massless and consequently move at the speed of light. They are also spin-½ particles and are therefore characterized by a certain **helicity**, which describes whether they are spin-up or spin-down when measured along the axis defined by their motion. (If the particles were massive, we could always work in their rest frame where there is no motion, but if they are massless they're always moving at the speed of light no matter what we do.) Helicity is classified as "left" or "right," corresponding to whether you would point your left or right hand along the direction of motion and allow your curled fingers to indicate the spin. Weinberg and Salam posited that the original SU(2) gauge symmetry applied only to left-handed fermions, leaving right-handed ones unchanged. That goes for quarks as well as leptons.

Which is great, except that as we've already noted, in the real world, electrons do have mass. Happily, spontaneous symmetry breaking once again comes to the rescue. For any particular fermion ψ, we can write down a gauge-invariant interaction of the form $y\Phi\psi^*\psi$. There are three fields involved: ψ is a left-handed fermion, ψ^* is a right-handed anti-fermion, and Φ is the Higgs scalar. The numerical factor y is a constant called the **Yukawa coupling**, after Japanese physicist Hideki Yukawa. The Yukawa coupling is a dimensionless number, because fermions have dimension $[M]^{3/2}$, the Higgs is a scalar field with dimension $[M]$, and each term in the Lagrange density has dimension $[M]^4$. There will be a separate term of this type for each kind of quark and charged lepton.

When the Higgs gets its vacuum expectation value, we can write $\Phi \to \bar{\Phi} + H(x,t)$, where $\bar{\Phi}$ is the fixed expectation value and H is the dynamical Higgs boson field. Then let's look at what happens to the Higgs/fermion interaction:

$$y\Phi\psi^*\psi \to y\bar{\Phi}\psi^*\psi + yH\psi^*\psi. \qquad (10.8)$$

What was a single term has now become two. The second one, $yH\psi^*\psi$, is just an interaction term between the Higgs and a fermion/anti-fermion pair, and its strength is proportional to y. In the first one, though, $\bar{\Phi}$ is not a dynamical field, it's a fixed constant. So there are only two fields involved in that one, the fermion and its anti-fermion. Which means that this is simply a mass term for the fermion!* The expectation value $\bar{\Phi}$ has dimensions of mass, and the Yukawa coupling

---

* I am told that it is good book-writing practice not to have too many exclamation points. But this one is deserved.

is dimensionless, so the units work out correctly for the combination $y\bar{\Phi}$ to be a mass. Each fermion can have a different mass, since y will be different for each flavor of particle, but they are all proportional to the Higgs expectation value. And as a bonus, this yields a very strong prediction: that the coupling between the Higgs boson H and each fermion is proportional to the fermion's mass, since they are both proportional to y. This feature was useful when it came to searching for the Higgs experimentally.

The fact that the Standard Model treats left- and right-handed fermions differently is why, even though Dirac didn't have to think hard to write down a mass term for the electron in his original equation, these days we have to "explain" the origin of mass. Dirac, in the 1930s, didn't know about parity violation, so he just wrote down the mass. Nowadays we are sufficiently constrained that we have to work a bit harder to explain where the electron mass comes from, and ultimately we attribute it to the Higgs field pervading all of space. It's not that the very notion of "mass" requires a Higgs field, it's just that the restrictive structure of the Standard Model would forbid fermion masses if it weren't for the Higgs getting a vacuum-expectation value and breaking the SU(2)×U(1) symmetry. That scenario, hatched in the 1960s, has been spectacularly confirmed in a long series of experimental results from the 1970s all the way up to the discovery of the Higgs boson itself in 2012.

Truly good ideas are rare, and sometimes it's hard even to recognize one when you see it. When Weinberg published his paper in 1967 with the modest title "A Model of Leptons," nobody paid much attention to it, including Weinberg himself. In the figure we see the number of citations to Weinberg's paper per year. There was no interest at all in the first few years, then it jumps up starting in 1971, settles down a bit, and jumps up again in 2010. For a long time it was the most highly cited paper in theoretical physics. We know what happened

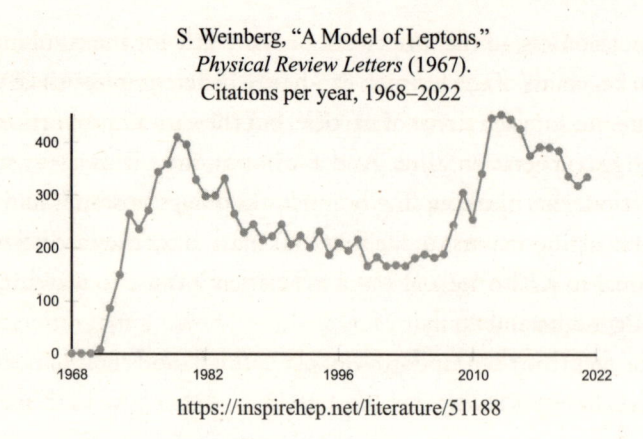

S. Weinberg, "A Model of Leptons,"
Physical Review Letters (1967).
Citations per year, 1968–2022

https://inspirehep.net/literature/51188

around 2010: the Large Hadron Collider started the search for the Higgs boson in earnest, eventually discovering it in 2012. What happened in 1971?

Recall that in those days the philosophy of effective field theories hadn't really taken hold, and a great amount of emphasis was put on whether a theory was renormalizable. And the thing was, in the case of Weinberg's theory, that was far from obvious. Weinberg himself tried to verify that it was renormalizable and got stuck; others didn't even try, since massive gauge bosons were thought to be a sign of nonrenormalizability.

What happened in 1971 was that Gerard 't Hooft and his PhD advisor Martinus Veltman showed that the theory is, in fact, renormalizable. Suddenly every particle physicist in the world was interested. It's an interesting case where scientists saw a dramatic leap in their confidence that a certain theory was on the right track, not because of any shocking experimental result but just from an improved theoretical understanding.

The electroweak theory plus the QCD model of strong interactions has an overall gauge symmetry of $SU(3) \times SU(2) \times U(1)$. Together with the specific set of fermionic particles (six quarks, six leptons) and of course the Higgs field, the overall theory is known by the underwhelming name of the **Standard Model of particle physics**. The finishing theoretical touches were put on the Standard Model in the 1970s, and experimentalists have continued to verify the existence of all of its various particles. The original version of the Standard Model set all neutrino masses to 0, but it's not hard to extend it to allow for massive neutrinos; the 2015 Nobel Prize in Physics was awarded to Takaaki Kajita and Arthur McDonald for showing that different flavors of neutrinos can undergo quantum oscillations into each other, which is only possible if they have mass. (The exact values of the neutrino masses have yet to be pinned down.) Other than that minor modification, the Standard Model has triumphantly passed every experimental test we have thrown at it to date.

ELEVEN

MATTER

The nuances of particle physics can seem somewhat distant from the concerns of our everyday lives. It's easy to forget that quarks and leptons and gauge bosons are inside all of us. We are literally made of them. In these final two chapters, we begin the journey that connects the physics within us to the world we usually see.

Here's an obvious question to ask: Why aren't atoms squishy? How is it possible to put together a collection of atoms in such a way as to make a solid object?

For a single atom, it's not too hard to understand why it maintains its shape, but we do have to break free of a common misconception. You will sometimes hear people say "atoms are mostly empty space." People have not been straight with you. That might have been true if you still believed in the Rutherford or Bohr models of the atom, with point-like electrons orbiting around the nucleus like a little solar system. And in that case, the solidity of atoms would indeed be hard to understand. But atoms are not empty space, at least not in our wave function–realist way of thinking. The electron wave function spreads out into a certain definite shape within each atom. You can try to deform that shape, but it will generally cost a lot of energy.

That helps explain the shape of a single atom, but how should we understand the fact that multiple atoms can *combine* to make solid objects? Why can't the electrons in two atoms just pile right on top of each other?

It's not, as people sometimes guess, because of the electromagnetic repulsion of the electrons. Back in 1842 Samuel Earnshaw proved a theorem that you can't trap charged particles, like electrons, in a stable configuration using only static electric fields, like those in a collection of atoms. Like most good theorems, you can try to wriggle out of the conclusions by denying some of the assumptions, but the intuition is on the right track. The solidity of matter is a subtler thing.

The real reason why matter is solid comes down to the fact that electrons are fermions, and fermions have a special property: no two of them can be in the same quantum state. That's why you can't pile them on top of each other. In this chapter we'll explore this feature and its connection to the spin of the particles involved.

BOSONS

Why do all electrons have the same mass and the same charge? It is not, as John Wheeler once suggested to Richard Feynman, because "they are all the same electron." Wheeler's idea came from the notion that positrons could be thought of as electrons moving backward in time. So an interaction in which an electron and a positron annihilate into a photon could be thought of as a single electron coming in, then "bouncing backward in time" to make the positron. This is actually not the right way to think about electrons, but it did help Richard Feynman invent Feynman diagrams.

The real reason why all electrons have the same mass and charge is that they are all excitations of a single underlying electron field. As a result, any two electrons are **identical particles**: they are

MATTER

indistinguishable from each other, even in principle. This, plus the magic of quantum entanglement, has crucial consequences for how they behave in nature.

Let's think about identical particles at the level of wave functions. Say we have two identical particles, and we use x_1 to denote the position variable for the first one, and x_2 for the second one. So we have an overall wave function $\Psi(x_1, x_2)$. Neither particle has "a location"; the variables x_1 and x_2 are two different labels on the same three-dimensional space where they could be observed. But if we do observe one of them, the fact that they are identical means there is no such thing as "which one" we observed. No physically observable quantities should change if we were to exchange x_1 with x_2.

There is an obvious way to make this happen: consider wave functions that are literally unchanged when we switch the two position variables. Such particles exist: they are the **bosons** we've been discussing for a while, such as the gauge bosons and the Higgs boson. They are named after Satyendra Nath Bose, who first studied their behavior in the 1920s. A wave function for two bosons is invariant under exchanging the particles with each other:

$$\text{Bosons: } \Psi_B(x_1, x_2) = \Psi_B(x_2, x_1). \tag{11.1}$$

When this is the case, we can think of two separate one-particle wave functions, ψ_1 and ψ_2, and entangle them in a symmetric fashion:

$$\Psi_B(x_1, x_2) = \psi_1(x_1)\,\psi_2(x_2) + \psi_1(x_2)\,\psi_2(x_1). \tag{11.2}$$

The same pattern extends to more than two particles in an obvious way.

There is no problem having two bosons in exactly the same quantum state. In (11.2) we simply set $\psi_1 = \psi_2$. Indeed, it turns out that bosons like being in the same quantum state; if there are bosons already in some state, it increases the probability for other bosons to transition into that state. This gives rise to important physical phenomena such as lasers and condensates, where large numbers of particles share the same low-energy quantum state. It also allows for the formation of macroscopic force fields, such as those we have in gravity and electromagnetism.

Bose, working at the University of Dhaka in a part of British India that is present-day Bangladesh, wrote a paper on the statistical properties of identical particles and sent it to Albert Einstein, hoping that he would arrange to have it translated from English to German for possible publication in *Zeitschrift für Physik*, one of the leading journals of the day. It was a slightly cheeky request, but not completely unwarranted; Bose had previously translated some of Einstein's papers into English for dissemination in India. Einstein was very impressed with Bose's paper; he did the translation himself, sent it to the journal under Bose's name, and wrote a companion paper that he submitted to the same journal. The statistical behavior of bosons, including their predilection for piling into the same quantum state, is today known as **Bose-Einstein statistics**.

FERMIONS

But electrons, with which we started this story, are not bosons at all. That's possible because in quantum mechanics, the requirement that exchanging identical particles should lead to the same physical properties doesn't necessarily mean that wave functions should simply be unchanged under particle exchange, as in (11.1). It is also possible that the wave function picks up a minus sign. That would leave the probability of observing the particle positions, given by the Born rule as $P(x_1, x_2) = |\Psi(x_1, x_2)|^2$, unaltered. There are indeed particles like this too; they are the **fermions**, named after Enrico Fermi.

$$\text{Fermions:} \quad \Psi_F(x_1, x_2) = -\Psi_F(x_2, x_1). \quad (11.3)$$

MATTER

We can do the same trick with fermions that we did with bosons, constructing the two-particle wave function by entangling two one-particle wave functions, but this time in an antisymmetric rather than a symmetric combination:

$$\Psi_F(x_1, x_2) = \psi_1(x_1)\psi_2(x_2) - \psi_1(x_2)\psi_2(x_1). \quad (11.4)$$

The "matter" particles of the Standard Model are all fermions. These include the six flavors of quarks (up, down, charm, strange, top, bottom), and six flavors of leptons (electron, muon, tau, electron neutrino, muon neutrino, and tau neutrino).

But now something remarkable happens. We mentioned that it was easy to put two bosons into the same wave function, and indeed that they preferred to be close to other bosons that way. Fermions are the opposite. If we try to set $\psi_1 = \psi_2$, we immediately see that (11.4) vanishes—it isn't a wave function at all. This is the famous **Pauli exclusion principle**, first posited by Wolfgang Pauli—no two fermions can occupy precisely the same quantum state.

Fermions, in other words, take up space. This is why they are generally associated with "matter," as opposed to the "forces" made of bosons. We now understand that they're all just quantum fields obeying different statistics. The properties of fermions are called **Fermi-Dirac statistics**, as Fermi and Dirac both studied them independently but Fermi reached his conclusions a little bit sooner.

Given that the minus sign in (11.3) is allowed because what we really care about is the probability $|\Psi(x_1, x_2)|^2$, you might wonder whether there are other kinds of particles whose wave functions transform as $\Psi(x_2, x_1) = e^{i\theta}\Psi(x_1, x_2)$ for some fixed parameter θ, since that would also leave the probability unaltered. The problem there is that the notion of "exchange two particles" has the property that if we exchange once and then exchange again, we want to get back to the original configuration. That will only happen if $e^{i\theta}$ is $+1$ or -1. In the special case where we are restricted to two spatial dimensions rather than the usual three, the rules are a bit looser, and a new kind of particle called **anyons** become allowed. They were hypothesized by Jon Leinaas and Jan Myrheim and colleagues in 1977, and their properties elucidated by Frank Wilczek in 1982. The fundamental particles of the Standard Model live in three dimensions and are strictly bosons or fermions, but there are materials that support collective excitations that effectively live in just two dimensions. The existence of anyons in such systems was only verified experimentally in 2020.

SPIN AND ROTATION

To emphasize, the properties of multi-particle wave functions—either they stay the same under exchange, or they pick up a minus sign—are what define particles as being either bosons or fermions. But you will often hear it said that bosons are particles whose intrinsic spin is an integer—0, 1, 2, and so on—while fermions are particles with half-integral spin—1/2, 3/2, 5/2, and so on. What is the relationship between these two ideas?

The relationship is that we can prove, within the framework of relativistic quantum field theory, that integer-spin particles are always going to be bosons, and half-integer-spin particles are always going to be fermions. It's a result, not a definition: the **spin-statistics theorem**. Despite the absolute centrality of this theorem to the success of quantum field theory, its proof is a bit mathematically delicate. Almost no modern textbook explains the argument, pointing instead to the classic 1964 text *PCT, Spin and Statistics, and All That*, by Ray Streater and Arthur Wightman. We're not going to prove it rigorously here either, but we can see why such a result might be plausible.

To get there, let's think a bit more about what "spin" really means. For a classical object, spin is simply a rate of rotation around some axis.

MATTER

The laws of nature feature a symmetry to the effect that we can rotate our frame of reference and the underlying laws are unchanged. Noether's theorem then implies the existence of a conserved quantity, and this quantity is the angular momentum of the object. Unlike classical objects, quantum particles can have an intrinsic spin that never changes its total amount, only its orientation in space. The electron is spin-½, the photon is spin-1, and so on. (We're using units where $\hbar = 1$, so the angular momentum of an electron is actually $\hbar/2$.)

Now we want to extend this to quantum fields. We begin by asking, what happens to a field when we perform a rotation in space? In fact, let's zero in on what happens to the field at just a single point as we do a rotation around that point.

Your intuitive answer is likely to be "nothing at all," since a rotation around some particular point leaves that point unmoved. And that's true if we're thinking about scalar fields, which are simply numerical values at each point. That value doesn't change, so the field itself doesn't change no matter what angle we rotate by.

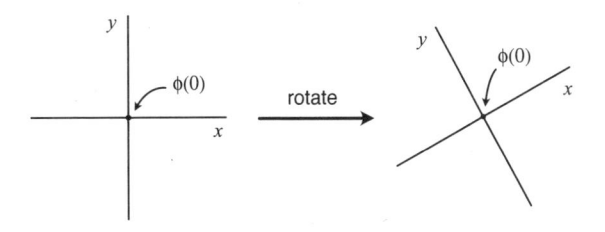

Once we move beyond scalar fields, things become trickier. Think of a vector field $\vec{v}$. At every point it has a direction in space as well as a magnitude. When we rotate around that point, the components of the vector change, unlike the value of a simple scalar. But there is a special case, namely when we rotate by exactly 2π radians (360 degrees). Then the vector does indeed come back to exactly where we started.

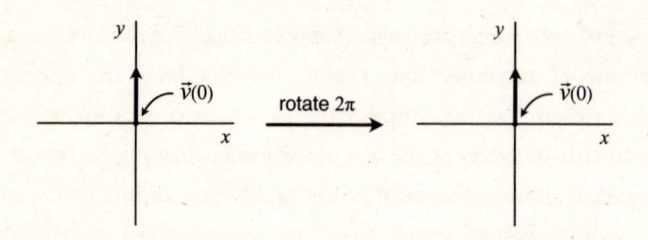

Could there be other examples where rotating by a certain amount returns us to where we started? Indeed there can be, and certain kinds of tensor fields qualify, including those describing gravitational waves (or gravitons, in the quantum language). Such a field has the property that it returns to its original state after a rotation of just π radians, halfway around the circle. If that seems hard to visualize, think of the tensor field h as a line segment with arrows on both ends, centered at the point we are considering. Then it becomes clear that a rotation by π returns the field at that point to its original value.

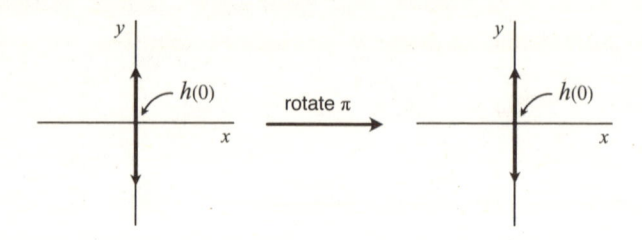

The big reveal is, these properties of fields under rotations are directly tied to the spin of the associated quantum particle. The scalar, spin-0, is invariant under any rotation. The vector, spin-1, is invariant under a 2π rotation. And the graviton, spin-2, is invariant under a π rotation. The general rule is:

A particle of spin s is invariant under a rotation by $2\pi/s$ radians.

We interpret that as "any rotation at all" when $s = 0$.

MATTER

CLASSICAL FIELDS AND SPIN

Reality checks are sometimes useful in these abstract realms of quantum field theory. Let's allow ourselves a brief digression to see how the spin behavior of photons and gravitons is reflected in the dynamics of classical electromagnetic and gravitational waves.

Electromagnetism first. A traveling wave consists of both an electric and a magnetic field, perpendicular to each other and oscillating up and down as they pass. Let's concentrate on the electric field and its effect on a proton as it passes by. At the proton's position, the passing electromagnetic wave looks like an electric field that oscillates upward, then down, then back again. The proton is accelerated in the direction of that field, so it will be nudged up and then back down in a repeating pattern.

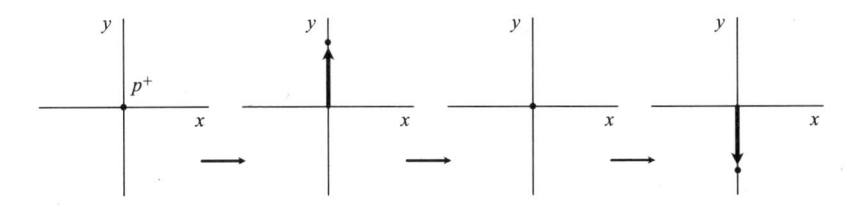

If we concentrate on the direction of the electric force on the proton, we see that at each moment of time, the pattern is invariant under a 2π rotation. That's just what we anticipated for a spin-1 particle, which the photon is.

Now contrast that with a gravitational wave. In this case we can't simply concentrate on a single particle. When we discussed the principle of equivalence in *Space, Time, and Motion*, we saw that a gravitational field is essentially undetectable at a single point in space, and that's just as true for a gravitational wave as it is for the gravitational field of a star or a planet. Given just one particle, we can always choose coordinates that stay centered on the particle so it doesn't seem to "accelerate" at all. Instead, we need to look at a collection of nearby particles and consider their relative motions, which we can think of as tidal forces from the passing gravitational wave.

So consider a ring of particles floating in space as a gravitational wave goes by. What happens is that the wave first stretches the configuration in one direction while squeezing it in the perpendicular direction, then relaxes back and stretches/squeezes in the opposite sense, continuing in an oscillating sequence.

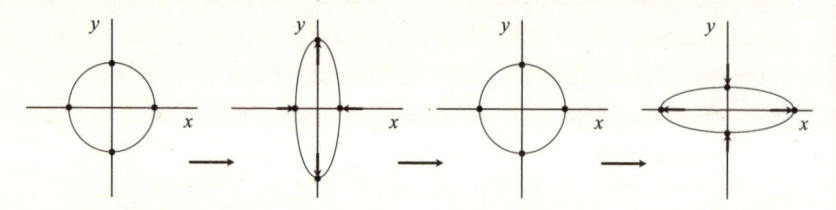

So when we look at the pattern of tidal forces at any one time, we notice that it is invariant under a rotation by π radians, rather than 2π as in the electromagnetic case. That's why the graviton is a spin-2 particle.

Two things to note about this simple demonstration. First, you might detect a family resemblance between the picture of the forces due to a passing gravitational wave and a possible oscillation of a circular loop of string. That's no coincidence. The reason why string theory, which was originally envisioned as an approach to the strong interactions, always ends up as a theory of gravity is that this particular mode of a vibrating closed string acts as a massless spin-2 particle, which we interpret as the graviton. (There's more to being a graviton than just having the right spin—most important, coupling to all forms of energy-momentum, to satisfy the equivalence principle—but string theory predicts those properties as well.)

The other thing is to think about gravitational-wave observatories, such as the LIGO facility at Hanford, Washington. It consists of two 4-kilometer-long arms extending at a right angle from a central facility. Those arms are evacuated tubes down which lasers are fired, to bounce off mirrors at the other ends and return, where changes in the distance traveled can be measured to fantastic precision. Now you know the reason why you need two arms, and why it's efficient to put them at a right angle to each other: because a passing gravitational wave will stretch things in one direction and squeeze them in the perpendicular one.

SPIN-½

Back to quantum field theory. We gave a rule according to which a particle of spin s is invariant under a rotation by $2\pi/s$ radians. The electron is a spin-½ particle, so it should be invariant under a rotation by 4π, which is twice around a circle. What kind of monstrosity is not invariant when we rotate it by 2π but is invariant when we rotate by 4π?

It's not as hard to imagine such things as you might think at first. Consider a Möbius strip, a simple band where two ends are connected with a half-twist between them. If you could imagine walking along a Möbius strip, after going around once you would find yourself on the opposite side from where you had started. It's only after walking around twice that you come back to your original orientation.

Another example that sticks closer to the notion of a "rotation" is holding a cube (or a coffee cup, or whatever) in your hand and rotating it by twisting your arm without tilting the object along any other axis. You can preserve the vertical orientation of the cube by moving it, for example, under your arm and then back. Your arm will be twisted a bit—please do try this at home, but please don't injure yourself—and it should be clear that there's no way to untwist it without moving the cube. But we can keep rotating the cube in the same direction—not backward, that would be cheating—this time moving over your arm instead of under. Somewhat miraculously, the cube and your arm will return to their original configuration, after having been rotated by 4π radians. The cube, considered in isolation, is invariant under rotations by 2π. But the relationship between the cube and the outside world, here represented by your arm, is only invariant under 4π rotations. (This move is part of the Balinese candle dance but was popularized among physicists in a lecture by Richard Feynman.)

The technical subject under consideration is known as the **representation theory** of the rotation group SO(3). The group itself is an abstract set of transformations; what actually happens to a particular object under such a transformation is the "representation" under which it transforms. Scalars don't change at all; vectors return to their configuration after a rotation of 2π; the particular representation under which gravitons transform returns to its configuration after a rotation of π.

Electrons and other spin-½ particles (which includes all of the fundamental fermions discovered to date) transform under a representation that returns to its original configuration after a rotation of 4π. Such objects are called **spinors**.

If the electron field ψ_e is unaffected by a 4π rotation, what happens to it under a rotation of 2π, just once around the circle? Probably you can guess: it picks up a minus sign.

$$R(2\pi) \cdot \psi_e(x) = -\psi_e(x); \; R(4\pi) \cdot \psi_e(x) = \psi_e(x). \qquad (11.5)$$

That makes sense, since doing a 2π rotation twice should be equivalent to doing a 4π rotation once. It's also a little bit provocative. When we talked about fermions, their defining feature was picking up a minus sign when we interchanged two identical particles. Now with spin-½ particles, we see that they pick up a minus sign under a 2π rotation. Keep that in mind for a bit.

MEASURING SPIN

When we first encountered spin in Chapter 2, we mentioned the intriguing outcome of the Stern-Gerlach experiment, which sent a beam of spin-½ particles through a pinched magnetic field. Spinning particles act like little magnets in their own right, so passing through an external magnetic field deflects them, depending on how their spin is oriented. But we don't get a spray of deflections across various angles; what we see is particles either deflected upward by a fixed angle or downward by a fixed angle. We have measured the spin of the particle along the chosen axis, and the two possible measurement outcomes are either spin-up or spin-down. Spin measurements give quantized outcomes for the amount of angular momentum along the axis of measurement.

But not all particles are spin-½. What happens in more general circumstances?

There are still quantized outcomes, but the possibilities will depend on the spin of the particle. The rule is that the possible outcomes are separated by one unit of spin (that is, by $\hbar$), and that they extend from minus the spin to plus the spin. That means that for spin-½ the possible outcomes are −½ or +½ for the component of angular momentum along the measurement axis. For a spin-0 particle, there is no deflection. But for a spin-1 particle, there are three possible outcomes: −1, 0, and +1. It can be deflected downward, upward, or not at all. For a hypothetical spin-³⁄₂ particle there would be four possible outcomes: −³⁄₂, −½, +½, and +³⁄₂. For a spin-s particle there will be $2s+1$ possible outcomes, ranging from −s to +s by increments of 1.

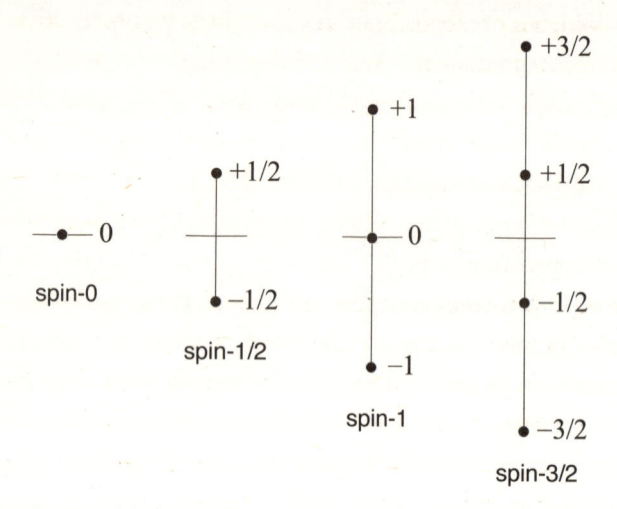

spin-0

spin-1/2

spin-1

spin-3/2

Massless particles are a special case, however. The $2s+1$ possible outcomes are what we could measure in the rest frame of the particle. But massless particles don't have a rest frame, since they're always moving at the speed of light. In that case, there are only two ways they can spin—either with their spin axis pointing along the direction of their motion, or opposite to it. Those are just the "helicity" states of massless particles we talked about in the previous chapter. So while a massive spin-1 boson has three possible spin states, a massless spin-1 particle (like a photon or gluon) has only two.

This explains a lingering puzzle from the previous chapter when

MATTER

we talked about the Higgs mechanism. When a gauge symmetry is spontaneously broken, rather than getting massless Goldstone bosons like we do when a global symmetry is broken, instead the gauge bosons become massive. The would-be Goldstone bosons have been "eaten" by the gauge bosons. But they don't disappear entirely, since a massive spin-1 gauge boson has three possible spin states rather than just two. The degree of freedom that would have been a separate spinless boson is now the spin-0 component of the massive gauge boson. There are still the same number of things that can happen in the world (varieties of vibrating fields); they just arrange themselves differently after spontaneous symmetry breaking.

THE SPIN-STATISTICS THEOREM

We've seen that exchanging two identical fermions multiplies the wave function by a minus sign, while rotating a single spin-½ particle by 2π also picks up a minus sign. The essence of the spin-statistics theorem—bosons all have integer spins, fermions have half-integer spins—is that these two minus signs are secretly the same minus sign. For the wave function of several identical particles, exchanging two of them is equivalent to rotating just one of them. Since fermions pick up a minus sign under interchange, it also has to be true that they pick up a minus sign when rotated by 2π.

We're not going to prove the theorem, which relies on details of relativistic quantum field theory, nor are we even going to sketch the reasoning behind the proof. Instead, we will simply explain how, in the spirit of the rotating-a-cube-on-your-arm example we looked at, it makes perfect sense that rotation and exchange can be equivalent, even though they are physically distinct operations.

Consider two cubes, with a ribbon loosely connecting them. The cubes are supposed to be identical particles, but we've shaded them differently so you can see when they are moved. Starting from a configuration without any twists in the ribbon, let's rotate one of the cubes by 2π, keeping the other cube fixed. To be definite, turn it in a clockwise direction as seen from above. There is now a twist in the ribbon; it is in a topologically different configuration from what we started with.

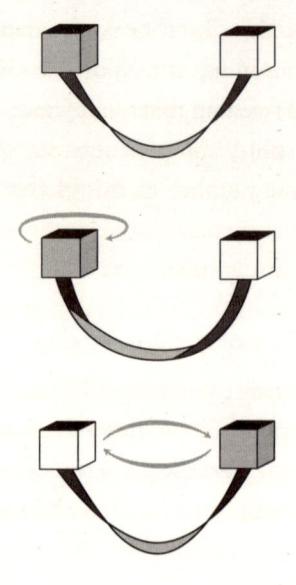

Now let's exchange the two cubes with each other, again in a clockwise sense, as viewed from above. We keep the cubes in a constant orientation along the way, and at a fixed height, so no extra rotations are happening. But it's clear that the ribbon has returned to its original configuration. Exchanging the two particles has undone the twist that we introduced by rotating just one of them. That's how fermions behave, at least in a relativistic universe with three dimensions of space.

MATTER
The fact that electrons are fermions explains why matter is solid. Because electrons obey Fermi-Dirac statistics, we know that no two of them can occupy the same quantum state—the Pauli exclusion

principle. But it took physicists quite a while to establish the connection carefully. The basic ideas were worked out by mathematical physicist Elliott Lieb in the 1970s.

To be fair, the conclusion isn't obvious. The exclusion principle doesn't say that two particles can't be in the same *place*, it says they can't be in the same *quantum state*. With two hydrogen atoms, we might be able to imagine overlapping the atoms so that their electrons are almost in the same state, but not quite. But quantum mechanics isn't going to let you get away with such trickery. The real demand of the exclusion principle is that any two electrons have to be in completely orthogonal quantum states. If you try to make two of them overlap in almost the same configuration, one or both will be pushed into a higher-energy state. This creates a repulsion between the electrons known as an **exchange force** or **Pauli repulsion**. It is this repulsive force that is responsible for atoms taking up space, and thus for the solidity of matter. (For bosons there is an analogous attractive force. If electrons had been bosons, the universe would have been an utterly different place.)

You might hesitate at the use of the word "force," since it's traditional to identify just four fundamental forces: gravitation, electromagnetism, and the strong and weak nuclear forces. But that identification was always casual and suggestive, not meant to be rigorous. As we've said, everything is really just quantum fields and their interactions. The repulsion between identical fermions is just part of the dynamics of a quantum field. It becomes useful to think of it as a "force" when we're letting our hair down and talking as if the world were made of classical particles, which is perfectly okay in appropriate circumstances.

In most ordinary materials, the exclusion principle explains why atoms take up space, but details of the individual atoms and how they join together to form molecules also play an important role in the structural properties of any given substance. Of course, there are also extreme circumstances. When a typical star uses up its available nuclear fuel and is no longer supported by the heat generated by fusion, it sheds its outer layers and the remnant core settles down to a state known as a **white dwarf**. The matter in a white dwarf is extremely dense, about 200,000 times denser than the Earth. The star is supported by **degeneracy pressure** caused by the fact that all of its electrons are packed as tightly as can be, given that they are fermions. It's the electrons that count, not the atomic nuclei, since the former are lighter and therefore have larger Compton wavelengths.

If the mass of the degenerate star exceeds the **Chandrasekhar limit**, protons and electrons will combine to make neutrons (emitting neutrinos in the process). Neutrons are also fermions, but massive and therefore with a smaller Compton wavelength, so we are left with an even smaller and denser **neutron star**. A white dwarf with the same mass as the sun will be roughly the size of the Earth, while a neutron star of the same mass will be just slightly larger than the city of Baltimore.

At even higher masses not even the neutron pressure can withstand the pull of gravity. There are no known configurations of ordinary matter that have a higher density than that, so the star will collapse to form a black hole. That doesn't violate the exclusion principle; some particles will just be forced into higher-energy states during the collapse, until ultimately they hit the singularity and quantum field theory is no longer up to the task of describing what goes on.

TWELVE

ATOMS

The word "atom" comes to us from the ancient Greeks. In the fifth century BCE, Democritus (and his teacher, Leucippus, whose writings are lost) argued that matter cannot be subdivided indefinitely and therefore must be made of some kind of basic indivisible particles. The concept was taken up by eighteenth-century chemists, who observed that the ratios of substances in certain chemical reactions could be best accounted for by positing the existence of elementary building blocks out of which all elements are made. They were right, as far as chemical elements are concerned, but physicists soon came to understand that atoms can be broken down into even smaller particles: protons, neutrons, and electrons, with the protons and nucleons further consisting of up quarks, down quarks, and gluons. We now think of elementary particles, which in turn are excitations in quantum fields, as the basic building blocks of matter.

It is nevertheless clear, and certainly important for the arrangement of things in the universe, that the world of our immediate experience features certain combinations of elementary particles and not others. Why are electrons and up quarks so common, and muons and top quarks so rare? And given the basic ingredients of electrons and up/down quarks, why do they assemble into the particular configurations that we actually observe?

THE CORE THEORY

Let's first be a bit more systematic about what the known quantum fields and associated particles actually are. Physicists often speak of the "Standard Model," which doesn't include gravity, since the Standard Model is a renormalizable quantum field theory, and the quantum version of general relativity would not be renormalizable. Having come this far in the book, we know that what matters is not renormalizability but the existence of an effective field theory for particles and processes below some ultraviolet cutoff. And if that's what we're interested in, there is no trouble including gravity. We can take the spacetime metric $g_{\mu\nu}$ and write it as the sum of a background metric plus a small perturbation, $g_{\mu\nu} = g^0_{\mu\nu} + h_{\mu\nu}$, where $g^0_{\mu\nu}$ is the metric of flat Minkowski spacetime. Then the perturbation $h_{\mu\nu}$ can be treated as an ordinary field propagating in flat spacetime as long as we stay away from black holes and other extreme gravitational phenomena. Its associated particle is of course the graviton. The resulting model has been dubbed the **Core Theory** by Frank Wilczek.

Many physicists have proposed clever geometric designs to help us visualize all of the particles in the Core Theory. The reality is, it's kind of a mess. That's okay; nobody expects it to be the ultimate beautiful Theory of Everything, if there is any such thing. Not only does the Core Theory not describe black holes and the Big Bang but it also doesn't have room for the dark matter that makes up most of the matter in the universe, and there are various unexplained fine-tunings and naturalness issues that continue to plague it. Its main virtue is that it fits the data, which is a considerable virtue indeed. So let's just embrace the mess and consider all of the particles in the Core Theory in turn.

ATOMS

The figure shows these particles, highlighting three basic properties: the electric charge, the mass, and whether they are fermions or bosons. All of the bosons are spin-1 except for the Higgs *H*, which is spin-0, and the graviton *h*, which is spin-2. All of the fermions are spin-½. We're only plotting particles and not their antiparticles, which involves some arbitrary choices; for example, we've shown the negatively charged *W* boson but not its positively charged counterpart.

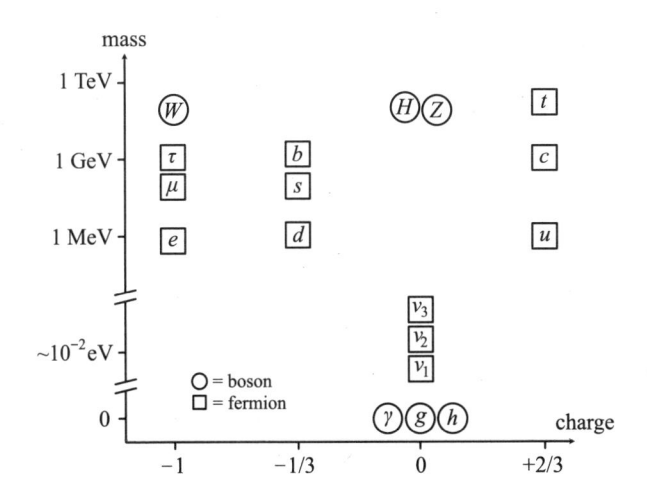

It's an unruly but ultimately manageable collection. Among the bosons we have the Higgs *H* (mass 125 GeV); the *W* (80 GeV) and *Z* (91 GeV) weak bosons; and the massless photon γ, graviton *h*, and eight gluons *g*. All of them are electrically neutral except for the *W*, with charge −1. There are two types of fermions, the strongly interacting quarks and the not-strongly-interacting leptons, and there are two subsets within each category, differing by one unit of charge. Different particles within the same subset are called different "flavors." So we have the down *d* (4.7 MeV); strange *s* (96 MeV); and bottom *b* (4.2 GeV) quarks, with charge −⅓; and the up *u* (2.2 MeV); charm *c* (1.3 GeV); and top *t* (173 GeV) quarks, all with charge +⅔. Charged leptons include the electron *e* (0.51 MeV); muon *μ* (106 MeV); and tau τ (1.8 GeV), all with charge -1. Then there are three neutrinos ν, all of which are, of course, neutral. Usually you will hear the three types of neutrinos associated with the charged leptons: there are electron neutrinos, muon neutrinos, and tau neutrinos, indicating which kind of neutrinos are created in weak interactions. But neutrinos are slippery beasts, and these three types are actually combinations of more basic flavors ν_1, ν_2, and ν_3. Each of the latter has a definite mass, which we know has to be less than 1 eV, but current experiments haven't pinned down precise values for the neutrino masses.

Finally, we have the rule that strongly interacting particles are confined into colorless combinations, the hadrons. The two major types are the mesons, consisting of one quark and one anti-quark, and the baryons, consisting of three quarks. It has been speculated that there could be bound states made purely of gluons—dubbed **glueballs** (don't blame me for the name)—but if they exist, they are going to be very short-lived.

WHAT REMAINS

Our universe is expanding, so long ago it was much denser and correspondingly hotter. As a rule of thumb, if the temperature is higher than the mass of a certain particle, that particle will be continually created and annihilated as the result of random thermal interactions. (Temperature and energy have the same units, and both are equivalent to mass when we set $c = 1$, so it's okay to compare the mass of a particle to the temperature of the stuff around it.)

The very early universe was therefore populated by all of the known particles in a high-temperature plasma, which we can think of as a gas that has become so hot that atoms can't survive, leaving only individual free particles. As the plasma expands and cools, the temperature drops below the masses of various particle types. At that point the particles can annihilate but there's not enough energy in a typical collision to create them anymore, and they drop off the scene.

We won't follow that process in detail but rather consider what happens to individual particles when they are left to evolve on their own.

We have the following three guidelines:

1) Heavier particles decay into lighter ones. In special relativity, energy is conserved but "mass" is not separately conserved; it is just one form of energy. A heavy particle can decay into a collection of particles with lower total mass, with the remaining energy being the kinetic energy of the offspring particles. There can also be absorption, where several particles convert into just one, but that's much rarer than decay. The underlying rules look the same forward or backward in time, so this must be because of increasing entropy. All else being equal, one particle has lower entropy than a collection of several particles.

2) Conserved quantities don't go away. For us the relevant quantities are going to be electric charge, lepton number (the number of leptons minus the number of anti-leptons), and baryon number (one-third of quark number, the number of quarks minus the number of anti-quarks).

3) Some particles interact and stick together into composite systems like atoms and nuclei, whereas others largely pass right through each other. Gravitons and neutrinos are passing through your body right now, but the probability of them interacting with you is negligibly small. (Dark matter particles might also be passing through you, but in the Biggest Ideas we're mostly sticking to things we know.)

Combining the first two of these rules, we can expect to find a lot of electrons, which are the lightest particles with electric charge; neutrinos, which are the lightest particles with lepton number; and protons, which are the lightest particles with baryon number. And precisely those particle types are quite common in our universe. But details matter, including why there are still neutrons in the world.

One additional cosmological fact worth mentioning is the matter/antimatter asymmetry. As far as the known laws of physics are concerned, there's no reason why there should be more atoms made of matter than made of antimatter, but there clearly are in the real world. We even know that other stars and galaxies are made of matter rather than antimatter, because otherwise we would expect high-energy radiation from the intergalactic regions between them, where matter and antimatter would gradually but surely come into contact and annihilate. We're not going to explain this here—since the explanation is unknown—but just take it for granted. One important detail is that we don't know for sure that there is more matter than antimatter, since there could be a bath of excess anti-neutrinos all around us. What we do know is that there are more baryons than anti-baryons, since there are no nearly invisible particles in which to hide excess anti-baryon number. For this reason the puzzle is known among physicists as the **baryon asymmetry**, and mechanisms invented to bring it about fall under the rubric of **baryogenesis**. The asymmetry is not that big; for every 10 billion anti-baryons in the early universe, there were 10 billion and one baryons. That extra one accounts for all of the baryons we see today.

DECAY MODES

Let's go through the various ways that particles can decay. As is our wont, we aren't going to be completely systematic, just hitting the central ideas. Our task is made simpler by the fact that photons, gravitons, and neutrinos don't have anything lighter to decay into. So there are four categories of things to worry about: the Higgs, the W/Z bosons, the quarks, and the charged leptons.

Start with the Higgs. It's a neutral, spinless particle, so its basic decay strategy will be to convert into a particle/antiparticle pair. That guarantees that all conserved quantities will be 0. Remember from Chapter 10 that the mass terms for all the other Standard Model particles originate in their coupling to the Higgs, and in (10.8) we saw that the fermion masses would be proportional to those couplings. Therefore, the Higgs wants to decay into the heaviest particles that it can—the relevant coupling constant is the same quantity that determines the particle's mass. (Decay into a top and anti-top is suppressed because even just one top quark is heavier than the Higgs.*) So the Higgs is happy to decay into a tau/anti-tau pair, and even happier to decay into a bottom/anti-bottom pair. You might worry about the latter possibility, since quarks are supposed to be confined into hadrons. That's no problem; because the strong interactions are strong, it's easy enough to create some more quarks and gluons along the way, in a process called **hadronization**. What we actually observe is the Higgs decaying into two mesons, or more realistically into a shower of mesons and baryons plus other miscellaneous photons and leptons that are created as by-products.

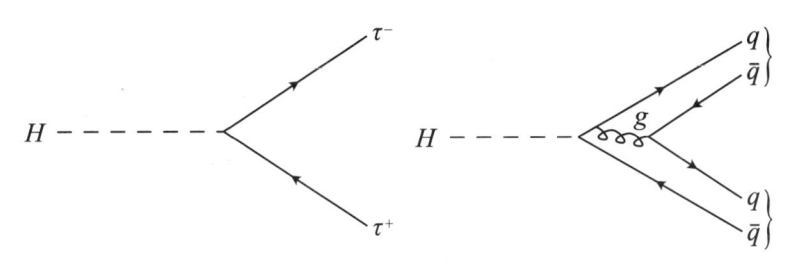

The W and Z bosons are going to decay in different ways, simply

---

* Suppressed, not completely forbidden, because the Higgs could convert into a *virtual* top/anti-top pair, as long as those virtual particles quickly decayed into other real particles. Virtual particles don't obey the usual relationship between rest mass and energy. But the further they are away from that usual relation, the less likely they are to be produced, so Higgs decay via top/anti-top pairs is still very rare.

because the *W* is electrically charged, and that charge has to go somewhere. A typical *W* decay is going to be into a fermion and an antifermion with charges differing by 1, such as an electron and an anti-neutrino or a down quark and an anti-up quark (which would then hadronize). The *Z*, meanwhile, decays much like the Higgs, into particles and their antiparticles, although at different rates.

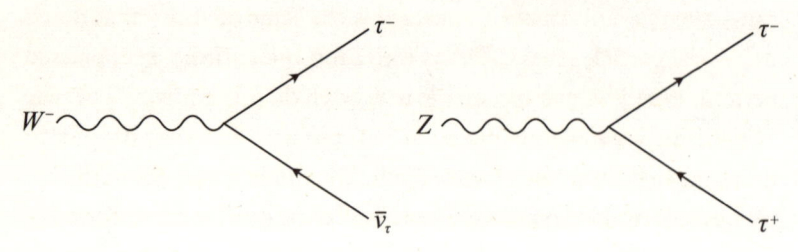

The *W* bosons, in fact, are the only particles in the Core Theory that have a fundamental vertex involving two distinct flavors of fermions. Photons, gravitons, gluons, the Higgs, and the *Z* all have vertices where they convert into a particle and its precise antiparticle, but not into a particle and a different flavor of antiparticle. Such processes would be **flavor-changing neutral currents**, since the decaying particles are all electrically neutral. As far as we know they never happen directly in nature (that is, there is no fundamental Feynman diagram vertex of that sort), but they can happen via loop diagrams, albeit at a smaller rate than we would expect for a direct interaction.

When it comes to fermions, then, their decay modes generally proceed via the emission of a *W* boson (a virtual one, for the lighter fermions) and conversion to a different flavor. The *W* then decays into another fermion and an anti-fermion. A muon, for example, generally decays by converting into a neutrino and a *W*, the latter of which can decay into an electron and its anti-neutrino. The same idea holds for quarks, even if they are inside a hadron; neutrons decay when one of their down quarks converts into an up quark and a *W*. This is even a common way for mesons to decay. A π^- meson, for example, is a bound state of a down quark and an anti-up. Those two can annihilate into a W^-, which then decays into an electron and an anti-neutrino.

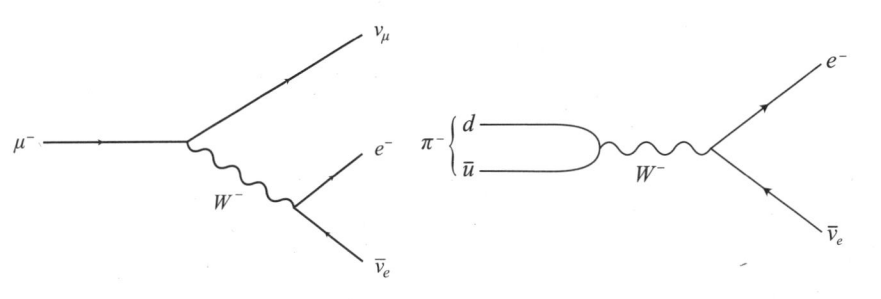

Because there are no direct flavor-changing neutral currents, heavier neutrinos cannot decay directly into lighter ones. We can imagine loop diagrams that would do the trick, but the associated rates would be extremely small and have never been observed in experiment. For all intents and purposes we can treat all of the neutrinos as stable particles.

With all of these allowed decays, given any initial collection of particles (with some imbalance of baryons over anti-baryons), we would expect it to quickly relax to a combination of just the following types:

- Massless unconfined gauge bosons, the photon and graviton
- Three types of neutrinos
- Electrons
- Up and down quarks confined into protons, the lightest particles with nonzero baryon number

That's a fairly good match for the universe we see, with one important exception: neutrons. There are neutrons all around us; almost half the mass of a typical human comes from the neutrons in their atomic nuclei. We have to think a bit harder to understand why they don't just decay away.

NEUTRONS AND NUCLIDES

Heavier particles tend to decay into lighter ones, given the constraints of conservation laws. The mass of a neutron is 939.6 MeV. It can, and does, decay into a combination of a proton (938.3 MeV), which carries away the baryon number; an electron (0.51 MeV), which compensates for the electric charge of the proton, and an anti-neutrino (< 1 eV), which compensates for the lepton number of the electron. The conserved quantities remain constant through the transformation, and the combined mass of proton + electron + neutrino is less than that of a single neutron.

But, we have to admit, barely. The neutron isn't that much heavier than the proton/electron/anti-neutrino troika, as can be seen by considering the fractional mass difference:

$$\frac{m_n - (m_p + m_e + m_{\bar{\nu}})}{m_n} \approx 0.001. \tag{12.1}$$

Intuitively, there isn't a lot of "room" for a neutron decay: the kinetic energies of the created particles have to be quite small. The mathematics of the decay rate reflects this, and neutrons end up living a long time (about fifteen minutes) compared to other unstable particles, which typically decay in tiny fractions of a second.

If the fundamental parameters of particle physics had been just a little bit different, protons could have been heavier than neutrons and would decay into them, rather than the other way around. That would make for a very boring universe. Our universe is interestingly complex because there are many kinds of stable nuclei, which are positively charged and can capture electrons, leading to rich forms of chemistry. If nuclei were all neutrons, they would be electrically neutral and unable to capture electrons. There would be no atoms, no chemistry, and no life.

But why are there nuclei at all? Why don't the neutrons inside them decay within a few minutes, rather than lasting essentially forever?

In Chapter 7 we talked about the binding energy of an electron to a proton—the mass of a bound hydrogen atom is less than the sum of the masses of the electron and proton, because there is a net negative energy in their electromagnetic attraction. If that seems weird, it's just a way of saying that we would have to inject energy to tear the atom apart.

The same idea applies to neutrons and protons in nuclei. The strong force is attractive, and neutrons and protons like to stick together. That means the energy of a nucleus can be less than the sum of its constituent protons and neutrons. For example, the simplest compound nucleus is the **deuteron** (nucleus of heavy hydrogen), which is just one proton and one neutron. It has a mass of 1,875.6 MeV, which can be thought of as the proton mass plus the neutron mass minus 2.2 McV of binding energy. If the neutron converted into a proton, electron, and anti-neutrino and the particles went their separate ways, the total mass would have to go up to 1,877.1 MeV. You can't decay into a heavier collection of particles. So the binding energy renders the deuteron stable, even though a free neutron is not.

A particular type of nucleus—specified by its number of protons and number of neutrons—is called a **nuclide**. The strong force wants to bind nucleons together, but protons have a positive electrical charge, which leads to mutual repulsion. As a result, there is a limitation on how big a nuclide can get. There are about 250 effectively stable nuclides—either perfectly stable, or with lifetimes greater than the age of the universe—and many more unstable ones.

[Chart: Neutrons vs Protons for nuclides, with markers H, C, Fe, Ag, Pb, U along top. black = stable nuclides, gray = unstable nuclides]

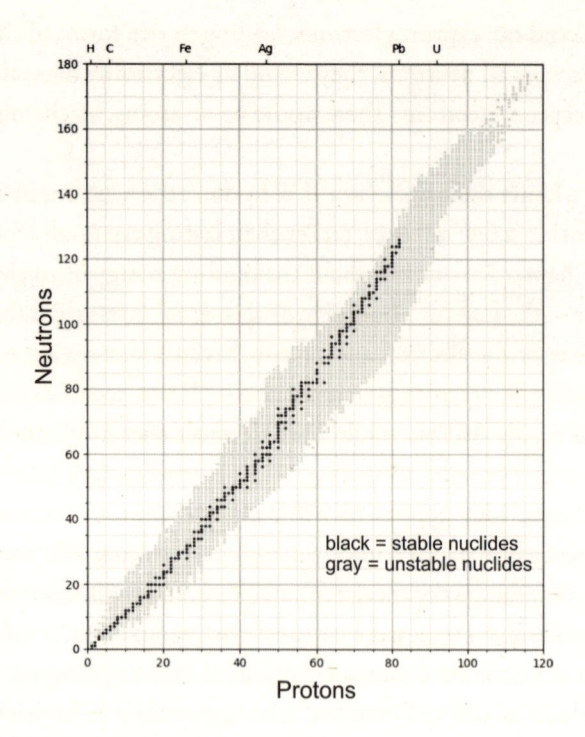

THE EVERYDAY WORLD

Once everything settles down, we're in a position to describe the basic ingredients of you and me and everything we see directly around us. Forget about neutrinos, which are out there in the universe but don't affect our everyday lives. For that matter, forget about the strong and weak nuclear forces; the strong force holds nuclei together, but once we know it's doing that, there's not much of a role for the strong force in the everyday world, unless our everyday lives are spent working at a nuclear power plant. (The nuclear forces play a crucial role in the fusion processes powering the sun, so in that sense they're important for our existence, but we don't need to keep them in mind when describing how we ourselves behave.)

What we're left with includes two kinds of particles making up "matter":

- Electrons
- 250 stable nuclides

And we also have two long-range forces that are relevant to macroscopic physics:

- Electromagnetism
- Gravity

That's basically it. You and all your friends and loved ones, not to mention your casual acquaintances and sworn enemies, are combinations of these matter particles interacting with each other and these forces according to the rules of quantum field theory. Every book you've ever read, every meal you've ever eaten, every painting you've seen and song you've heard have arisen out of the collective dynamics of these basic pieces. It's an amazing intellectual accomplishment for human beings to have figured all this out. There is still an enormous amount of work to be done to show how these microscopic constituents come together to produce the wonders of the macroscopic world, but the list of ingredients is well understood.

These ingredients don't have equal roles to play. Among the forces there is a crucial distinction: gravity is simple-minded, while electromagnetism is clever. Gravity is quite subtle when we start thinking about cosmology and black holes, but here on Earth it's straightforward: everything pulls on everything else. That's it. This is why there is no realistic hope of creating an antigravity device or gravity-manipulation machinery; gravity moves only in one direction. This is in dramatic contrast with electromagnetism, where we have both positive and negative charges. This simple fact allows us to cancel electromagnetic fields, to focus them, and generally to manipulate them with pinpoint precision. Electromagnetism can both push and pull, which gives it an enormously greater variety of effects than we get with gravity.

Likewise, electrons and nuclei play very different roles. The list of nuclides provides variety: we have a collection of chemical elements we can string together in marvelous molecular combinations. But given those ingredients, it's the electrons that are doing the interesting work, simply because they are much lighter and therefore easier to manipulate. It's electromagnetically interacting electrons that get pushed around to create electricity, that get shared among atoms to create chemistry, and generally that power the processes of technology and life.

ELEMENTS

Opposite charges attract, so nuclei and electrons like to team up to make atoms. Electrons are fermions, subject to the Pauli exclusion principle. This makes atoms both complicated and interesting.

Hydrogen atoms have just one electron, which will settle into a unique ground state if it's allowed to. Deuterium (one proton and one neutron) is heavier than ordinary hydrogen, but it counts as the same element because it has the same number of protons and therefore is atomically very similar. The next heaviest element we can imagine constructing from our ingredients is helium, with two protons and two neutrons in the nucleus, so two electrons in a neutral atom. According to the exclusion principle, those two electrons cannot occupy the same quantum state. But there is another variable lurking around: the spin of the electron, which can be either up or down. The two electrons in a helium atom can therefore have exactly the same spatial profile, as long as they have opposite spins. However, we cannot fit any more electrons into the same spatial wave function, or **orbital** as they like to say in chemistry. This is why helium is a **noble gas**—each atom has a full complement of electrons in its outermost (and in this case only) orbital, so it doesn't feel any need to gain or lose or share with other atoms and thus remains free of molecular commitments.

The next two elements are lithium and beryllium, with three and four protons, respectively. In each of them, there are two electrons snuggled up to the nucleus much as in helium, but also more electrons lurking in higher-energy states beyond that initial orbital. You might therefore expect lithium, with a single electron outside the filled first orbital, to be somewhat hydrogen-like in its chemical properties, and that is correct.* You might also expect beryllium to be a noble gas like helium, but that is completely wrong.

The reason why is that there is only one spatial configuration for the lowest-energy orbitals, but there are multiple possible configurations for the next-highest-energy orbitals. In fact, there are four of them: one that is spherically symmetric but which oscillates radially from the origin on out, and three that are not spherically symmetric but rather extended along one of the three spatial axes. So after helium, we can fit $2 \times 4 = 8$ new electrons in these four orbitals before we fill everything up. The next noble gas is in fact neon, with ten protons.

And so on, with more complications along the way as we go through heavier elements and more allowed electron orbitals. The specific ways that electrons can surround nuclei while remaining compatible with the exclusion principle are responsible for why atoms fall into the pattern described by the **Periodic Table** of the elements. It is all very intricate and beautiful, and you should be glad that this series of books concentrates on fundamental physics rather than chemistry and materials and other collective phenomena, or it would be a hundred books long.

---

* Persnickety chemists are tearing their hair out at casual statements like this; the outermost electron in lithium is much less tightly bound than that in hydrogen, so there are important chemical differences between the two. But we've already outraged persnickety physicists, mathematicians, and philosophers, so it's fair to spread things out.

CHEMISTRY

The fun doesn't end there. From a physics perspective, when we construct a list of stable nuclides or the atoms in the Periodic Table, what we're really doing is looking for locally stable configurations of Core Theory fields. A configuration is "locally stable" if it has lower mass than any other configuration it could easily transform into. Generally that means that either it's truly stable, or the only way it could change is to pass through other configurations that have higher mass. That might happen if an appropriate nudge is applied. A candle sitting on a table is locally stable, but if you light it, it will start burning, transforming into a set of combustion products with lower mass overall and higher total entropy.

Atoms aren't the only locally stable configurations of nuclides and electrons; we also have molecules. Chemists distinguish between a number of different types of chemical bonds, but the basic idea is that we can find lower-energy configurations of electrons orbiting more than one nucleus. Consider two hydrogen atoms, which can combine to make a hydrogen molecule H_2. The electrons (with opposite spins, to keep Pauli happy) can settle into a mutual orbital around both protons. The resulting energy is about 4.5 eV lower than two hydrogen atoms separately, with the difference representing the binding energy of the molecule.

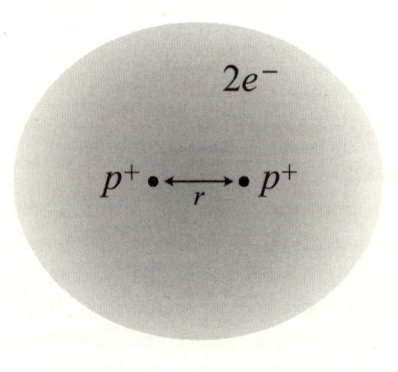

In this configuration, the protons will be separated by some distance r. Closer in, and the electrostatic repulsion between the protons would push them apart; farther away, and the electrons would tug them closer. We can think of this in terms of a potential energy of the system as a function of the separation. The binding energy is just this potential energy, evaluated at its minimum.

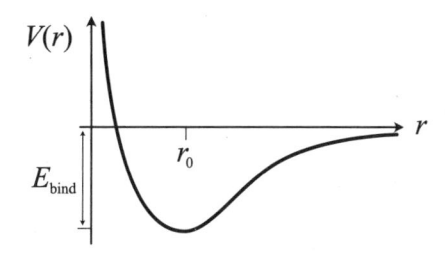

We started this book by thinking about quantized emission of light from atoms, when an electron jumps from one orbital to another. In molecules, there will once again be excited energy levels over and above the ground state; as we mentioned in Chapter 7, these have characteristic energies of order 10^{-2} eV. And of course there are all sorts of more complicated molecules that can be made, interacting in various ways under different conditions. In the end it's all just applied electromagnetism in a world with a couple hundred nuclides and the Pauli exclusion principle, but such a world is incredibly rich indeed.

THE PHYSICS UNDERLYING THE EVERYDAY WORLD

The Core Theory is obviously not the final theory of physics. It doesn't describe dark matter or conditions of strong gravitational fields, and it features various coincidences and fine-tunings that suggest a more complete explanation yet to come. People have had some ideas about what such a complete theory might be like, but as of right now we don't know. Nor do we know how close we are; a paper might be published with the final answer tomorrow, or we might still be looking for it a thousand years from now.

But there is a regime in which the Core Theory is entirely successful. If we don't have strong gravitational fields, and we only consider processes taking place well below some prudently chosen ultraviolet cutoff (say 10 GeV), the basic structure of quantum field theory assures us that the Core Theory is the complete story. We can certainly imagine other fields that we haven't yet discovered—the vast majority of physicists would bet there are plenty out there—but they are either so massive that they never get produced in interactions below the cutoff, or they couple to ordinary matter so weakly that they can be completely ignored.*

One of the reasons we can make such sweeping statements with such confidence is a feature of QFT known as **crossing symmetry**. This is not a symmetry of physical fields but of the Feynman diagrams we use to describe their interactions. Roughly it says we can rotate any diagram by 90 degrees—converting particles to antiparticles when their direction in time gets reversed—and get another diagram with an equal amplitude.

So let's imagine we had some new hypothetical fermion X, which we claimed interacts with electrons via exchange of a new hypothetical boson Y. Crossing symmetry says that the existence of an interaction $X + e^- \to X + e^-$ implies the existence of a related interaction $e^+ + e^- \to X + \bar{X}$. So if X interacts noticeably with electrons, we can easily create X in the lab by colliding electrons with positrons and watching what comes out. Happily, colliding particles and watching

---

* The axion is an example of a conjectured particle that has a low mass but is very weakly coupled. It could very well be the dark matter and is passing through your body right now, but it interacts so rarely that it has no effect on you whatsoever. That's why physicists have to build huge, expensive experiments just to have any hope of detecting it.

what comes out is particle physicists' favorite thing to do. We have extremely good data on what gets produced by such processes, and no mysterious particles have been found. Any particle that isn't already in the Core Theory would have to have a very weak coupling indeed.

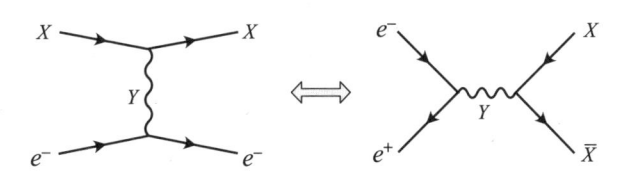

The remarkable consequence of this line of reasoning is that we have excellent reason to believe that the laws of physics underlying everyday life are completely known. There absolutely could be new particles and forces out there, but they don't interact noticeably with the protons, neutrons, and electrons of which we are made. And needless to say there is enormous work to be done to understand the rules governing the collective behavior of many such particles. Knowing the Core Theory doesn't help us much with biology or psychology, or for that matter aeronautics or climate science. It barely helps us with chemistry. That's the power of effective theories: we don't need to know everything that's going on at small distances and high energies in order to usefully talk about long distances and low energies. But in the everyday-life regime, we do happen to know all the ingredients and relevant dynamics of those microscopic ingredients.

The figure on the next page summarizes the situation. Solid arrows indicate forms of dependence that are definitely there, and dashed arrows are additional possibilities. The everyday-life regime is part of a larger macroscopic level that includes the rest of the universe, not all of which is completely understood (to say the least). At the level of quantum field theory, we have the Core Theory and possible other particles and forces. Astrophysics may very well depend on all of those, but you and I and the world we interact with depend only on

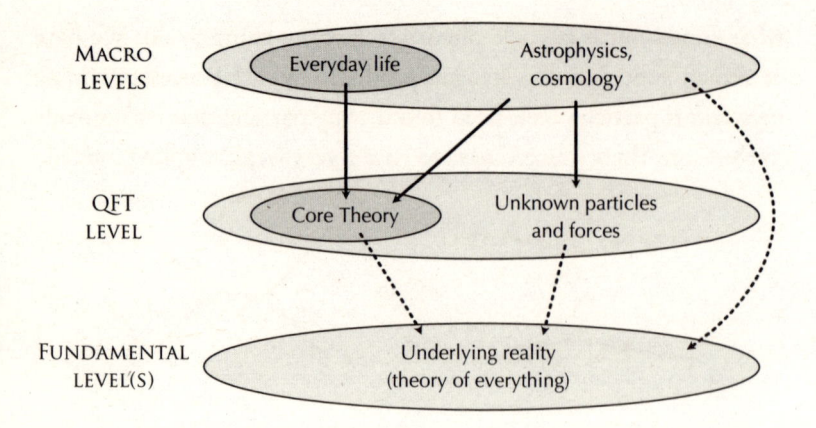

[Diagram: Three levels shown as ellipses.
- MACRO LEVELS: "Everyday life", "Astrophysics, cosmology"
- QFT LEVEL: "Core Theory", "Unknown particles and forces"
- FUNDAMENTAL LEVEL(S): "Underlying reality (theory of everything)"
Arrows connect Everyday life and Astrophysics/cosmology down to Core Theory; dashed arrows connect Core Theory and Unknown particles/forces down to Underlying reality; dashed curve connects Astrophysics/cosmology directly to Underlying reality.]

the Core Theory. And there might be even deeper levels, perhaps where spacetime itself is emergent. In some sense we depend on what happens at that level as well, but only insofar as it intersects with the Core Theory. We don't need to know any of the specifics to understand protons and neutrons and electrons and how they come together to form the world around us.

There are plenty of possible loopholes in this line of reasoning. Maybe the fundamental tenets of quantum field theory are somehow wrong. That would be surprising indeed, as QFT is the robust, low-energy limit of any theory based on quantum mechanics, relativity, and locality. Nevertheless, as good scientists we have to admit that anything is possible. For that matter, quantum mechanics itself might go wrong somehow.

But I wouldn't bet on any of those possibilities. We should always admit that we don't know everything but not be so cautious that we don't want to admit when we *do* know things. We know the particles we are made of, and the forces by which we interact. That's one of the greatest accomplishments in the history of humankind. We can be a little bit proud.

APPENDIX

FOURIER TRANSFORMS

Wave phenomena pose a special mathematical challenge for the physicist. The classical state of a single particle, or any finite collection of particles, is fully specified by a finite list of numbers: in three dimensions, the three components of position and three components of momentum, for each particle. Whereas a wave is specified, in principle, by an infinite number of numbers: the value of the wave at each point in space. In certain special cases we might be lucky enough to deal with wave profiles that can be neatly specified by a particular special function, but in general we will be overwhelmed with information in even the simplest case.

And the real world is full of waves. Even classically, electromagnetism convinced physicists once and for all of the centrality of fields, and waves within them, to describe the world. Once quantum mechanics comes along, the wave function for what classically would have been a single particle is now a function of space and time. And we soon realized that the thing to be quantizing isn't particles but rather fields. Modern physics is a story of waves upon waves.

We might think, "Okay, let's just zoom in on one point in space (or configuration space, in quantum mechanics) and ask what happens there, and then knit different points together, calculus-style." A noble idea, but not really practical. The behavior of a field typically depends not only on its value at any one point but on its derivatives in space and time. In practice, this means that a field can be doing one thing at one moment and something radically different the next. We can all visualize a calm sea that suddenly witnesses a roaring wave come through.

And yet, somehow, we human beings deal with waves all the time. If you are reading this book, your sensory and nervous systems are taking light waves and converting them into useful information in your brain. If you are listening to the audiobook version, an analogous thing is happening, but with sound waves.

We make use of sound and light waves in various ways, including pinpointing the directions from which they come. But let's focus now on our ability to recognize colors of light and pitches of sound. In both cases we're picking out the frequency, or wavelength, of the relevant wave. They are related by $f = v/\lambda$, where f is the frequency, λ is the wavelength, and v is the velocity of the wave. What's really impressive is that we can do that, at least to some degree, even when a bunch of different frequencies are superimposed on top of each other. You do that every time you pick out different voices singing in harmony.

Our brains do this kind of processing subconsciously, but it raises an interesting mathematical question: given a profile of some kind of wave, which might look messy and not at all regular, is there any systematic way to pick out specific frequencies?

The answer is yes, and the technique to do it is the **Fourier transform**. It is named after French mathematician and physicist Joseph Fourier, who in 1822 published an influential treatise on the theory of heat flow between objects. There he speculated that any function can be expressed as a sum of purely sinusoidal (sine-like or cosine-like) functions, although perhaps you would need an infinite number of

APPENDIX

them. In the figure we give a simple example of a messy-looking function decomposed into a sum of just four sinusoidal functions with different heights. Note that the heights of the waves on the right side of the figure are not all equal—the Fourier transform trades information about the value of the function at every point for information about the height of each wave into which it is decomposed. The dramatic idea is that this works, and the decomposition is perfectly unique and calculable, for any starting function at all.

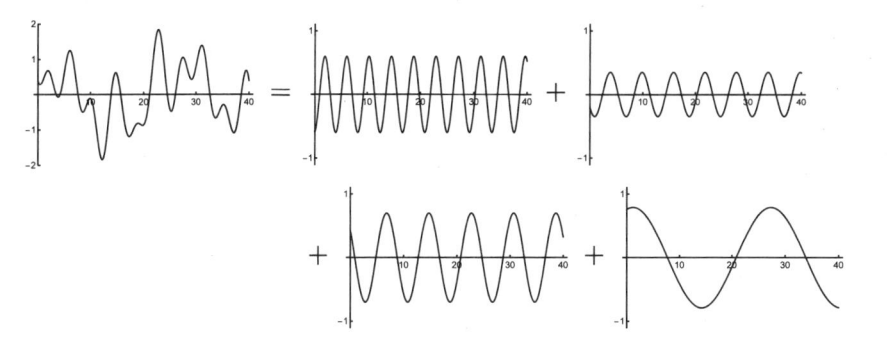

When I first learned about Fourier transforms as a college sophomore, I had no idea what use they would be and thought of them as a classic example of mathematical formalism for its own sake. I couldn't have been more wrong: in modern physics, it's hard to think of a more useful or ubiquitous operation than the Fourier transform. They appear everywhere waves do, which is just about everywhere.

Let's be a little more specific. Say we have some function of a single variable, $f(x)$. The claim is that this function can be written as a sum of sinusoidal functions (potentially an infinite number of them). Each of these sinusoidal functions is going to have some wavelength λ. For various reasons it's more convenient to use the **wave number** k, which is 2π times the number of wavelengths per distance:

$$k = 2\pi/\lambda. \qquad (A.1)$$

Then what we're saying is that the information contained in the original function $f(x)$ is equivalently contained in a list of contributions from sine waves of all possible wave numbers k. Let's call the contribution of some particular wave $\tilde{f}(k)$. The set of all such contributions is a function itself, but it's a function of k rather than the original variable x. And the basic idea is that these two functions contain the same information:

$$f(x) \leftrightarrow \tilde{f}(k). \tag{A.2}$$

The new function is known as the "Fourier transform" of the original. The real complication along the way is that everything is more convenient if we use complex numbers.

Let's just directly state the formulas for calculating the Fourier transform, and also for going backward to the original function:

$$\tilde{f}(k) = \frac{1}{\sqrt{2\pi}} \int f(x) e^{-ikx} \, dx, \tag{A.3a}$$

$$f(x) = \frac{1}{\sqrt{2\pi}} \int \tilde{f}(k) e^{ikx} \, dk. \tag{A.3b}$$

Notice the minus-sign difference in the two exponentials, and that we integrate over x to get a function of k and vice versa. Also note that these formulas are specific to one-dimensional functions; the factors of 2π in front will be different in different numbers of dimensions. Finally, different references will use different conventions in various ways. If you're doing a calculation on which your life (or grade) depends, make sure to check that your conventions are consistent.

The exponentials of imaginary numbers are a nice way of dealing with sinusoidal functions, using the following lovely fact:

APPENDIX

$$e^{i\theta} = \cos\theta + i\sin\theta. \tag{A.4}$$

Here, $\cos\theta$ and $\sin\theta$ are the familiar cosine and sine trigonometric functions, and $e = 2.718\ldots$ is Euler's number. The cosine starts at one and initially goes down, while sine starts at zero and initially goes up, both oscillating with a period 2π. We met the function e^x in *Space, Time, and Motion* because of a special property of its derivative: $\frac{d}{dx}e^x = e^x$. Now we find another fun property of this function, its relationship to trigonometry when the exponent is an imaginary variable rather than a real one. Equation (A.4) is known as Euler's formula, after Swiss polymath Leonhard Euler. When we set $\theta = \pi$, we get $e^{i\pi} = \cos\pi + i\sin\pi = -1 + 0$, which is equivalent to the famous expression

$$e^{i\pi} + 1 = 0. \tag{A.5}$$

There's a good case to be made that 0, 1, i, π, and e are the five most important numbers in all of mathematics. This formula relates them in a stylish way.

What is really going on in the inscrutable formulas (A.3a–b)? In Chapter 4 we pointed out that Hilbert space, which we first encountered as the space of all possible wave functions, can be thought of as a vector space. You can add wave functions together and multiply them by complex numbers, and the result is a different wave function. But for a one-dimensional particle with wave function $\Psi(x)$, the corresponding Hilbert space is infinite-dimensional. We can think of every value of x as representing a different dimension of the Hilbert space, and the value $\Psi(x)$ is the component of the infinite-dimensional vector along that axis.

When you have a vector space, you are free to change the basis you use, with a corresponding change of components. The figure below shows a two-dimensional vector space with (x, y) coordinates, on which we've also indicated different coordinates (s, t), related to the originals by

$$s = \frac{1}{\sqrt{2}}(x+y), \quad t = \frac{1}{\sqrt{2}}(x-y). \tag{A.6}$$

The factors of $1/\sqrt{2}$ enable a nicely symmetric inverse transformation, as you can check:

$$x = \frac{1}{\sqrt{2}}(s+t), \quad y = \frac{1}{\sqrt{2}}(s-t). \tag{A.7}$$

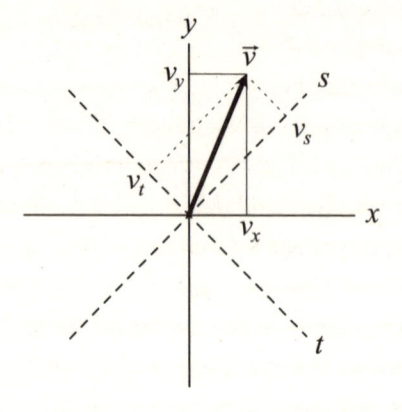

Any given vector $\vec{v}$ can be decomposed equally well in the (x, y) basis or the (s, t) basis, as a sum of appropriate components times basis vectors:

$$\vec{v} = v_x \vec{e}_x + v_y \vec{e}_y = v_s \vec{e}_s + v_t \vec{e}_t. \tag{A.8}$$

The components in different bases are related in exactly the same way the coordinates are. For example:

$$v_s = \frac{1}{\sqrt{2}}(v_x + v_y), \quad v_t = \frac{1}{\sqrt{2}}(v_x - v_y). \tag{A.9}$$

In other words, the information contained in the vector, as encoded in the components in one basis, can be traded in for the components in the other basis, and vice versa:

$$\{v_x, v_y\} \leftrightarrow \{v_s, v_t\}. \tag{A.10}$$

If this looks similar to (A.2), that's entirely intentional. A Fourier transform is just a change of basis, but in an infinite-dimensional vector space. We speak of the "position basis" and the "momentum basis," recalling that in quantum field theory wave number and momentum are related by $p = \hbar k$.

Fourier transforms show up in physics whenever it is useful to think of functions as sums of waves with different heights and wavelengths rather than as values at each point in space. There are many such circumstances. One is the Heisenberg uncertainty principle, which states that there are no wave functions that are simultaneously localized in both position and momentum. Once we know that the momentum-space wave function is simply the Fourier transform of the position-space wave function, this makes perfect sense. We have rotated the basis, just as we did in a two-dimensional vector space in the above figure. It seems pretty clear that there is no vector pointing perfectly in the x direction (no y component) that is also pointing perfectly in the s or t directions. That is the heart of the uncertainty principle; you can't be simultaneously localized in terms of two different bases that are rotated with respect to each other. (The nontrivial physics content is the realization that the momentum wave function is not independent of the position wave function, but rather is the Fourier transform of it.) Generally speaking, if a function of x is highly peaked, its Fourier transform will be broadly spread out, and vice versa.

The other obvious example is turning a free quantum field theory into a collection of simple harmonic oscillators. Remember that there is a gradient energy term in the Lagrangian for a field, which looks like ½ $(\partial_x \phi)^2$. If we look at the formula (A.3b) for a position-space function in terms of its Fourier transform, on the right-hand side the only place that x appears is within the exponential, e^{ikx}. And the derivative of that is $\partial_x e^{ikx} = ik e^{ikx}$. So taking a partial derivative with respect to x can be thought of, once we've Fourier-transformed everything, as

$$\frac{\partial}{\partial x} \to ik. \tag{A.11}$$

We have traded in a derivative for multiplication by a number. Instead of worrying about solving differential equations, which can be hard, we can worry about solving an algebraic equation, which is much easier.

Physically, the upshot is that a free scalar field is a collection of simple harmonic oscillators, one for each wave number, and we know how to solve that. Upon quantization, the energy levels of those oscillators turn into numbers of particles. Fourier transforms explain where the quanta come from in quantum field theory.

INDEX

Abel, Niels Henrik, 180
abelian groups, 180, 186, 216
absorption, 102
alpha particles, 10
amplitudes, 45, 51, 110–13
Anderson, Carl, 101, 208
Anderson, Philip, 224
antimatter, 101
anti-neutrino, 101
antiparticles, 105–7
anti-quarks, 160
antisymmetry, 204
anyons, 240
Aspect, Alain, 75
associativity, 176n
asymptotic freedom, 142, 218
atoms. *See also* electrons; natural units; neutrons; protons
    constituents of, 149
    core theory, 254–56
    hydrogen, 161–63, 266
    overview, 9–10, 253
    shape of, 235–36
Avogadro's number, 153

Bardeen, William, 216
baryogenesis, 258
baryon asymmetry, 258
baryon number, 159–60
baryons, 159–60, 256
Bell, John, 74
Bell's theorem, 74–75, 75n
beryllium, 267
beta decay, 101, 102, 227
bilateral symmetry, 169
binding energies, 159, 163, 164, 263, 268–69
bit, definition of, 50–51
blackbody radiation, 12–15, 124
Bohmian mechanics, 74
Bohm's nonlocal theory, 74, 75
Bohr, Niels, 17, 18
Bohr model, 19
Bohr radius, 162–63
Boltzmann, Ludwig, 15
Born, Max, 20–21, 37
Born rule, 37
Bose, Satyendra Nath, 237–38
Bose-Einstein statistics, 238

bosons, 97–98, 146, 236–38, 239. *See also* gauge bosons; Goldstone bosons; Higgs bosons; W bosons; Z bosons
branches, 73
bras, 46n
Brout, Robert, 224

Chandrasekhar limit, 252
chemistry, 268–69
circle symmetries, 181–83
classical fields, 79–80
classical mechanics
    entanglement, 62
    Hamiltonian operator in, 24–26
    Lagrangian defined in, 113
    Laplacian paradigm of, 33
    matter, 18
    measurement in, 34
    momentum, 51–52
    phase space, 44
    symmetries in, 167
    wave functions, 38
classical physics, 2
Clauser, John, 75
closure, 176n
color charges, 158–59, 216, 217
color space, 194–95, 197, 199–200, 215
complex conjugation, 188, 191n
complex numbers, 22, 187–92, 276. *See also* real numbers
complex plane, 22, 188–89
complex vector space, 189–91
Compton, Arthur, 154–55
Compton wavelength, 154–56, 158, 162
confined phase, 214
confinement, 158–59, 218–20
conjectured particles, 270n
connections, 197–200
constructive interference, 39
Core Theory, 254–56, 269–72
cosmological constant, 93n
cosmological constant problem, 146–47
Coulomb phase, 214, 220

coupling constants, 110–12, 135, 136–38, 141
crossing symmetry, 270–71

Dalton, John, 9
de Broglie, Louis, 19, 74, 154
de Broglie wavelength, 19–20, 53–54, 154
de Broglie–Bohm theory, 74
decay modes, 102, 258–62
decoherence, 70–72
degeneracy pressure, 252
Democritus, 253
destructive interference, 39
deuterium, 266
deuteron, 263
dihedral group of degree 3, 173, 175
dimensional analysis, 132, 134–35, 141, 145–46, 163–64
Dirac, Paul, 46, 101, 206–8, 239
discrete groups, 183
discrete symmetries, 168
distance, as inverse energies, 133–34
double-slit experiment, 39–41, 43, 72, 154
Dyson, Freeman, 103, 128

Eames, Charles and Ray, 149
Earnshaw, Samuel, 236
effective field theories
    definition of, 124
    hierarchy problem in, 146–47
    loop diagrams, 125–28
    overview, 124–25
    renormalization, 128–32, 139
effective Lagrangians, 139–40
eigenstates of spin, 28–29, 49
Einstein, Albert, 16, 36, 63, 74, 146, 238. *See also* EPR
electromagnetic fields, 100–101
electromagnetic radiation, 8
electromagnetism, 8–9, 16, 162–63, 203–4, 214–15, 243, 265–66
electron fields, 76, 118, 201–7, 210–12, 236, 247
electron volts, 150–51

INDEX

electrons
 discovery of, 10
 as fermions, 236
 as identical particles, 236–37
 matter, role in, 266
 obeying Fermi-Dirac statistics, 250
 orbital, 17–20, 266–67
 overview, 9–10
 as particles, 43
 Schrödinger's equation and, 30
electroweak theory, 138, 227–33
elementary particles, 253
energy, measuring, 150–51.
 See also momentum
energy density, 81, 87–90
energy eigenstates, 28–29
energy quanta of light, 16
energy-momentum conservation, 125–27
Englert, François, 224
entanglement
 in classical mechanics, 62
 example of, 58–59
 measurement apparatus and, 70–71
 as natural consequence of quantum mechanics, 59–62
 quantum measurements and, 67–70
 wave functions and, 21, 90
EPR
 element of reality, 66
 spooky action at a distance, 63–65
 z-spins, 66–67
equation of parallel transport, 198
Euler, Leonhard, 277
Euler's formula, 277
Everett, Hugh, III, 73
everyday life, laws of physics underlying, 271
exchange force, 251
excited states, 28, 92
exclusion principle. *See* Pauli exclusion principle

Faraday, Michael, 8
Fermi, Enrico, 101, 227, 238–39

Fermi-Dirac statistics, 239–40, 250
fermions, 97–98, 146, 157, 230–31, 236–40, 260
Feynman, Richard, 103–4, 128, 236
Feynman diagrams, 104–7, 112–13, 117–22
Feynman's path integral, 113–14
field configuration, 78, 85–86, 90–92
field theory, Lagrangian approach to, 113–17
fields
 bosons (*See* bosons)
 classical behavior of, 80–82
 definition of, 7
 derivatives of, 80–81
 fermions (*See* fermions)
 free fields, 83–85, 90, 92, 100–101
 gauge, 199–200
 interaction of, 98
 kinetic energy, 79–80, 94–95
 Lagrangians for, 113–17, 119, 206, 208–9
 mass of, 83
 particles from, 92–96, 99
 positron, 118, 210–11, 212
 scalar field (*See* scalar fields)
 wave functions of, 90–92
field-strength tensor, 204–5
Fifth Solvay Conference, 36
fine-tuning, 145
flux tubes, 219
Fock, Vladimir, 95
Fock space, 95, 97
folk theorem, 97
foundations of quantum mechanics, 72–76
Fourier, Joseph, 274–75
Fourier transforms
 calculation formulas, 276–77
 Heisenberg uncertainty principle, 279–80
 modes, 85–87
 overview, 53
 plane waves, 94
 purpose of, 274

Fourier transforms (*cont.*)
  simple harmonic oscillators, 280
  sinusoidal functions, 275–76
  as useful, 275–76
free fields, 83–85, 90, 92–93, 100–101, 115–16, 280
free-scalar field theory, 115–16
Fritzsch, Harald, 216

gauge bosons, 200, 214–15, 217n, 220, 223–29, 249
gauge fields, 199–200. *See also* fields
gauge invariance, 8n, 167, 200–207, 209–10, 212–14, 224–25
gauge symmetry, 196
gauge theories
  connections, 197–200
  for long-range forces, 208–10
  overview, 193–94
  phases of, 214
  quantum chromodynamics, 142, 158–59, 194, 214–17
gauge transformations, 196–97, 202
gauge-covariant derivatives, 202–3
gauge-invariant kinetic term, 225
Geiger, Hans, 10
Gell-Mann, Murray, 157, 215–16
Gerlach, Walter, 48
Glashow, Sheldon, 228
global symmetry transformations, 196
glueballs, 256
gluons, 10, 157, 200, 217, 256
Goldstone, Jeffrey, 223
Goldstone bosons, 223, 224, 226, 249
gradient energy, 81–82, 115, 116
Grand Unification, 152
gravitation, 220
gravitational waves, 243–44
gravitational-wave observatories, 244–45
gravitons. *See also* gauge bosons
  in core theory, 254–55
  in Coulomb phase, 214, 220
  decay mode, 260
  existence of, 98n

  human interaction of, 257
  as long-range forces, 210
  spin behavior of, 242, 243–44, 255
gravity, 7–8, 265
Greenberg, Otto, 215
Gross, David, 218
ground states, 28, 92
group axioms, 175–76
group theory, 167–68, 176–78
Guralnik, Gerald, 224

hadronization, 259
hadrons, 157, 219, 256
Hagen, Carl, 224
Hamiltonian operator, 24–27, 29
harmonic oscillators. *See* simple harmonic oscillators
Heisenberg, Werner, 20–21
Heisenberg uncertainty principle, 50, 54–56, 67, 109, 155, 279–80
helicity, 229–30
helium, 266–67
helium nuclei. *See* alpha particles
Hermann, Grete, 74
hidden-variable theories, 74–75
hierarchy problem, 145–47
Higgs, Peter, 224
Higgs bosons, 59–61, 63, 138, 145, 152, 208, 231–32
Higgs decay, 258–59
Higgs field, 229
Higgs mass, 145–46, 152
Higgs mechanism, 223–27
Higgs phase, 214
Hilbert, David, 45n
Hilbert space, 44–46, 62, 190, 277
Huygens, Christiaan, 16
hydrogen atoms, 161–63, 266

identical particles, 236–38, 247, 249–50
identity transformation, 172–73
imaginary numbers, 22, 86, 187–88, 276–77
infinities, 123–24, 128

INDEX

infrared (IR) phenomena, 13, 90, 130–32, 139–40, 147
integers, 178–81
integers modulo, 179–80
integral, definition of, 38
interaction vertices, 104–5
interactions. *See also* beta decay; electroweak theory
   free fields, 116–17
   Lagrangian describing, 116–17
   between particles, 102–5
   in quantum electrodynamics, 210–12
   between quarks and gluons, 158
   scalar fields, 138–39
   weak, 227
interference patterns
   in de Broglie's model, 19–20
   decoherence destroying, 72
   double-slit experiment, 39–41, 72, 154
internal vector space, 195
invariant objects, 169
irrelevant operators, 142
isomorphism, 177, 180

Jeans, James, 13
Jordan, Pascual, 20–21
joule, definition of, 14

Kajita, Takaaki, 233
Kelvin, William Thomson, Baron, 11
kets, 46, 47
Kibble, Tom, 224
kinetic energy. *See also* Lagrangians
   blackbody radiation and, 12
   electron volts, 150
   of everyday phenomena, 152–53
   of fields, 79–80, 94–95
   Newtonian formula for, 132
   overview, 79–82
   of particles, 81, 88–89

Lagrange density, 114–15, 134–35, 141
Lagrangians. *See also* effective Lagrangians
   definition of, 113

   describing interacting scalar fields, 116–17
   for fields, 113–17, 119, 206, 208–9
   as free field theory, 115–16
   gauge-invariant, 206, 212
Landau, Lev, 142
Landau pole, 142
Laplace, Pierre-Simon, 7
Laplacian paradigm, 33
Larmor, Joseph, 17
Lee, Tsung-Dao, 229
Leinaas, Jon, 240
Lepton number, 160
leptons, 157
Leucippus, 253
Leutwyler, Heinrich, 216
Lie, Sophus, 183
Lie groups, 183, 186
Lieb, Elliott, 251
light. *See also* electromagnetism
   Einstein's energy quanta of, 16
   Maxwell's theory, 1, 8–9
   particle-like properties of, 12
   photons, 8, 16–17
   wave nature of, 39–40, 274
lithium, 267
local symmetry, 196
locality, 64–65, 80–81, 97, 114, 272
loop diagrams, 125–28
loop momenta, 129–30
Lorenz-invariant, 82

magnetic fields, 8
manifold, 183
Many-Worlds interpretation of quantum mechanics, 73–74
marginal operators, 142
Marsden, Ernest, 10
massive objects, 56
massless fields, 8n
massless gauge bosons, 223–24, 226, 228, 249
massless particles, 120, 157, 194, 209–10, 214–16, 223, 248

matrix mechanics, 20–21
matter, 17–18, 250–53, 264–69
Maxwell, James Clerk, 8–9, 15, 16, 203–4, 205
McDonald, Arthur, 233
measurement. *See* quantum measurements
measurement apparatus, interacting with environment, 70–72
the measurement problem, 34, 42
mesons, 160, 256, 260–61
Mexican-hat potential, 221–22
Michelson, Albert, 11
Mills, Robert, 213
Möbius strip, 245
"A Model of Leptons" (Weinberg), 231–32
modes
  energy density, 87–90
  Fourier transforms, 85, 87
  with infinite wavelengths, 93
  oscillation frequency, 88–89
  overview, 84–87
  wave number, 85–86
  wave vector, 86
molecules, 164–65, 268–69
momentum
  in classical mechanics, 51–52
  de Broglie's equation, 53–54
  Feynman diagrams and, 120–22
  Heisenberg uncertainty principle, 54–56
  internal, 127
  inverse relationship with space/time, 129–30
  as observables, 20, 51–52
  of virtual particles, 125
momentum basis, 52–53
momentum eigenstate, 53
momentum four-vector, 125–27
momentum-space wave functions, 279
Moo-Young Han, 215
muon, 260
Myrheim, Jan, 240

Nagaoka, Hantaro, 17–18
Nambu, Yoichiro, 215, 223
Nambu-Goldstone boson, 223
natural units, 132–34, 150–52
naturalness principle, 144–47
neon, 267
neutral pions, 219
neutrinos, 151, 233, 256
neutron star, 252
neutrons, 9–10, 157, 160, 261–63
Newton, Isaac, 7, 16, 132
noble gases, 266–67
Noether's theorem, 167, 212, 241
non-abelian gauge theory, 199, 213
non-abelian groups, 180, 186
non-renormalizable theories, 137–38
nonzero energy, 93
normalized wave function, 38
nuclear forces, overview, 214–15
nucleons, 153
nuclides, 263, 266

objective collapse models, 75
observables, 20, 36, 51–52
orbital electrons, 17–20, 266–67
orthogonal groups. *See also* special orthogonal groups
  of a circle, 182–83
  vector spaces and, 184–87
oscillators. *See* simple harmonic oscillators

parallel transport, equation of, 198
parity, 172, 229
parity violations, 231
particle physics scales, 151–53
particles. *See also* antiparticles; spinning particles
  acting like waves, 17–20
  alpha, 10
  annihilation of, 96–97
  calculating, 104
  conjectured, 270n
  creation of, 96–97
  decay modes, 258–62

INDEX

electrons as, 43
evolution of, 257
from fields, 92–96, 99
interactions between, 102–5
kinetic energy of, 81, 88–89
massless, 120, 157, 194, 209–10, 214–16, 223, 248
overview, 9–11
populating the universe, 256–58
real vs. virtual, 119–20
scattering, 102–3
scattering probabilities, 109–10
size of, 154–56
as superposition of many locations, 47
as unstable, 60
virtual, 119–22
waves acting like, 15–17
path integral (Feynman), 113–14
Pauli, Wolfgang, 239
Pauli exclusion principle, 239, 250–52
Pauli repulsion, 251
Periodic Table of the elements, 267, 268
perturbation theory, 112–13
phase space, 44
phase(s), definition of, 213–14
photoelectric effect, 16–17
photon fields, 201, 203–4, 211
photons, 1, 8, 16–17, 208–9.
  See also gauge bosons
pilot wave theory, 74
Planck, Max, 13–15, 16
Planck blackbody radiation law, 14–15
Planck's constant, 14, 15, 19, 24
plane waves, 84, 94
plum pudding model, 10
Podolsky, Boris, 63
Politzer, David, 218
position, as observables, 20, 51–52
position eigenstate, 47
position-space wave functions, 279–80
positron fields, 118, 210–11, 212
positrons, 101, 207–8
potential energy, 79–80, 83, 116
Powers of Ten (film), 149

probabilities of events, 37–38
propagator diagrams, 117–18
protons, 9–10, 157

quanta, definition of, 1
quantum chromodynamics (QCD), 142, 158–59, 194, 214–17
quantum electrodynamics (QED)
  coupling constants, 111–12, 137–38
  Feynman diagrams for, 112–13, 118–20
  gauge invariance, 200–201
  interactions, 210–12
  Lagrangian for, 140
  overview, 108
  quantum chromodynamics vs., 215–17
quantum field theory (QFT), 3, 77–78, 100, 158
quantum foundations, epistemic approaches to, 75
quantum measurements, 33–34, 41–44, 59, 67–70, 75
quantum mechanics, 1–2, 20–23, 56, 67, 72–76
quantum objects, 2
quarks, 9–10, 157–61, 194–96, 215, 260
qubits, 48–51

Rayleigh, Robert John Strutt, Baron, 13
real numbers, 177. *See also* complex numbers
real particles, 119–20
realism, 64, 76
the reality problem, 42
reduced Planck constant, 14
reflection, 171
relativity, 82
relevant operators, 141–42
renormalization, 124, 128–32, 135–36, 137–39, 141–44, 218
representation theory, 245–46
Riemann tensor, 205
Rosen, Nathan, 63
rotations, 240–42

INDEX

running coupling constants, 141
Rutherford, Ernest, 10, 17

Salam, Abdus, 228–30
scalar field theory
  effective, 143–44
  with interactions, 138–39
  kinetic term in, 224–25
  with Lagrange density, 134–36
  renormalization, 141–44
scalar fields
  applying quantum mechanics rules to, 99
  definition of, 78
  with dimension, 230
  energy density of, 81
  gauge invariance and, 206–7, 224–25
  harmonic oscillators and, 85, 89–90
  Higgs mechanism, 224–27
  Lagrangian describing, 116–17
  symmetries and, 220–23
scaling argument, 162–63
scattering, 102–3
scattering amplitude, 104–5
Schrödinger, Erwin, 21–22, 35, 38–39
Schrödinger's equation, 23–27, 30, 72–73, 91–92
Schwinger, Julian, 103, 128, 228
simple harmonic oscillators
  formula for, 93n
  Fourier transforms, 280
  free fields, 83, 85, 92–93, 115–16, 280
  free scalar fields, 89–90
  molecules and, 164–65
  overview, 27–30
  quantum fields behaving as, 78–79
sinusoidal functions, 275–76
sound waves, 274
Space, Time, and Motion (Carroll), 3–4
spacetime, 129–30
special orthogonal groups, 183, 185–87. *See also* orthogonal groups
special unitary groups, 191–92
spectrum of blackbody, 12–13

sphere, 181–82, 184
spin behavior, of gravitons, 242, 243–44, 255
spin-½ particles, 49, 51, 245–47
spinning particles, 48–51, 63–64, 247–49
spinors, 201, 246–47
spin-statistics theorem, 240–42, 249–50
spontaneous symmetry breaking, 220–23, 224, 226, 230
spooky action at a distance, 64–65
Standard Model of particle physics, 4, 168–69, 231–33, 239–40
Stern, Otto, 48
Stern-Gerlach experiment, 48–50, 68, 247
string theory, 219–20, 244
Strutt, John (Lord Rayleigh), 13
SU(3) symmetry, 169, 195, 196, 215–16, 217, 228–30, 232
SU(3) transformations, 195, 196, 216
subgroups, 178–79
supersymmetry, 146
symmetries. *See also* spontaneous symmetry breaking; SU(3) symmetry
  as bilateral, 169
  breaking, 220–23
  circle, 181–83
  in classical mechanics, 167
  as discrete, 168
  group theory, 167–68
  groups of, 173–76
  matter/antimatter, 258
  Mexican-hat potential, 221–22
  object transformations, 169, 171–72
  rotations, 171
  sets of, 168
  triangles, 170–73
  of two-dimensional planes, 183–84

't Hooft, Gerard, 138, 232
tensor fields, 198, 204–5, 208–9, 242
thermal radiation. *See* blackbody radiation
Thompson, J. J., 10
Thomson, William (Lord Kelvin), 11
time, as inverse energies, 133–34

INDEX

time reversal, 107
Tomonaga, Shin'ichirō, 103, 128
transformation of an object, 169
tree diagrams, 127
triangle symmetries, 170–73

ultraviolet (UV) theory, 90, 130–31
ultraviolet catastrophe, 13
ultraviolet cutoff, 130–31, 139, 142, 145
uncertainty principle. *See* Heisenberg uncertainty principle
unitary groups, 168, 190–92
units. *See* natural units
universe, particles populating, 256–58
Utiyama, Ryoyu, 198

vacuum, definition of, 92–93
vacuum energy, 146
vacuum expectation value, 222–23
vacuum states, 92–93, 222
valence quarks, 161
vector potential, 199, 205
vector spaces, 44–47, 184–87, 189–90, 195, 277–79
Veltman, Martinus, 138, 232
vertex, 104–5
virtual particles, 119–22, 125
von Neumann, John, 45n, 74

W bosons, 193–94, 215, 227–28, 255, 258–60
Ward, John, 228
wave function collapse, 35–36
wave function of the universe, 57–58, 59–62
wave functions
  Born rule, 37
  in classical mechanics, 38
  coaxing particle-like behavior from, 35–36
  components of, 45
  with definite momentum, 54
  double-slit experiment, 39–40, 43
  energy levels for, 99–100
  of field configurations, 90–92
  as functions of configuration space, 62
  interference patterns, 39–41
  multi-particle, 240
  normalized, 38
  overview, 21–23
  representing reality, 76
  Schrödinger's equation and, 35
  as vectors, 46–47
wave mechanics, 21
wave packet, 94
wave vector, 86
wave-particle duality, 39–41
waves
  acting like particles, 15–17
  light, 39–40, 274
  particles acting like, 17–20
weak interactions theories, 227
Weinberg, Steven, 228–30, 231–32
Wheeler, John, 163, 236
white dwarf, 251–52
Wien, Wilhelm, 13
Wilczek, Frank, 218, 240, 254
Wilson, Ken, 128–30
Wu, Chien-Shiung, 229

Yang, Cheng Ning, 213, 229
Yang-Mills theories, 213, 224
Young, Thomas, 39–40
Yukawa, Hideki, 230
Yukawa coupling, 230–31

Z bosons, 193–94, 200, 258–60
Zeilinger, Anton, 75
zero energy, 93
zero total spin, 63
zero-sphere, 182
z-spins, 63, 66–67
Zweig, George, 157, 215

ALSO BY SEAN CARROLL

The Biggest Ideas in the Universe 1: Space, Time and Motion

The instant *New York Times* bestseller

Immense, strange and infinite, the world of modern physics often feels impenetrable to the undiscerning eye – a jumble of muons, gluons and quarks, impossible to explain without multiple degrees and a research position at CERN.

But it doesn't have to be this way.

Allow world-renowned theoretical physicist and bestselling author Sean Carroll to guide you through the biggest ideas in the universe. Elegant and simple, Carroll unravels a web of theory to get to the heart of the truths they represent about the world around us.

In *Space, Time and Motion*, the first book of this landmark trilogy, Carroll delves into the core of classical physics. From Euclid to Einstein, *Space, Time and Motion* explores the ideas which revolutionised science and forever changed our understanding of our place in the cosmos.